Springer Tracts in Advanced Robotics 116

Editors

Prof. Bruno Siciliano
Dipartimento di Ingegneria Elettrica
e Tecnologie dell'Informazione
Università degli Studi di Napoli
Federico II
Via Claudio 21, 80125 Napoli
Italy
E-mail: siciliano@unina.it

Prof. Oussama Khatib
Artificial Intelligence Laboratory
Department of Computer Science
Stanford University
Stanford, CA 94305-9010
USA
E-mail: khatib@cs.stanford.edu

More information about this series at http://www.springer.com/series/5208

Alexander Dietrich

Whole-Body Impedance Control of Wheeled Humanoid Robots

Springer

Alexander Dietrich
Institut für Robotik und Mechatronik
Deutsches Zentrum für Luft- und Raumfahrt
Wessling
Germany

ISSN 1610-7438 ISSN 1610-742X (electronic)
Springer Tracts in Advanced Robotics
ISBN 978-3-319-82129-0 ISBN 978-3-319-40557-5 (eBook)
DOI 10.1007/978-3-319-40557-5

Softcover reprint of the hardcover 1st edition 2016

Printed on acid-free paper

This Springer imprint is published by Springer Nature
The registered company is Springer International Publishing AG Switzerland

Series Foreword

Robotics is undergoing a major transformation in scope and dimension. From a largely dominant industrial focus, robotics is rapidly expanding into human environments and vigorously engaged in its new challenges. Interacting, assisting, serving, and exploring with humans, the emerging robots will increasingly touch people and their lives.

Beyond its impact on physical robots, the body of knowledge robotics has produced is revealing a much wider range of applications reaching across diverse research areas and scientific disciplines, such as biomechanics, haptics, neurosciences, virtual simulation, animation, surgery, and sensor networks among others. In return, the challenges of the new emerging areas are proving an abundant source of stimulation and insights for the field of robotics. It is indeed at the intersection of disciplines that the most striking advances happen.

The *Springer Tracts in Advanced Robotics* (STAR) is devoted to bringing to the research community the latest advances in the robotics field on the basis of their significance and quality. Through a wide and timely dissemination of critical research developments in robotics, our objective with this series is to promote more exchanges and collaborations among the researchers in the community and contribute to further advancements in this rapidly growing field.

The monograph by Alexander Dietrich presents the outcome of an extensive research work on whole-body impedance control of wheeled humanoid robots. The augmented task space approach is adopted to manage multiple task objectives and hierarchically handle priorities among them, while exploiting the available redundant degrees of freedom of the mobile manipulator through projection onto the null space of its Jacobian. Typical tasks include self-collision avoidance, singularity avoidance, mechanical joint limits, posture control, and impedance control. Stability of the system with the resulting multi-objective impedance controller is addressed. The approach is experimentally validated in a number of real-world service robotics applications.

Remarkably, the monograph is based on the author's doctoral thesis, which was a co-winner of the 2016 Georges Giralt Ph.D. Award. A very fine addition to STAR!

Naples, Italy
April 2016

Bruno Siciliano
STAR Editor

Preface

This monograph is based on my Ph.D. thesis written at the Institute of Robotics and Mechatronics of the German Aerospace Center (DLR) in Oberpfaffenhofen. The thesis was defended at the Technical University of Munich (TUM) in October 2015.

The book addresses the topic of whole-body impedance control of wheeled humanoid robots. Such systems are usually characterized by a large number of actuated degrees of freedom, which can be utilized to execute several subtasks at the same time. One example is to manipulate an object while simultaneously moving through a room in an energy-efficient way, avoiding collisions with the environment, and preventing the arms from reaching singular configurations. The goal of the present research is to design a control task hierarchy to achieve all of these objectives following a given order of priority. In other words, tasks with lower priority must not disturb tasks with higher priority and will be performed in the best possible way under this restriction. In the present work, several control tasks are developed and integrated into such a hierarchical framework. Besides experimental validations on a mobile humanoid robot, the first formal proof of stability for a hierarchical torque controller is presented. While several aspects treated in this book consider wheeled systems in particular, most of the concepts are deliberately presented in a generic way and can be used on mobile systems with other locomotion principles as well, for example on legged humanoid robots, underwater vehicles, or flying systems. Household chores from the field of service robotics are chosen as benchmark to experimentally validate the presented approach in relevant real-world applications.

During my research, many people accompanied me with advice and technological support. In particular, I would like to express my deep gratitude to my supervisor and mentor Prof. Alin Albu-Schäffer for his guidance and the inspiring discussions we had throughout the course of this work. Furthermore, my special thanks go to Dr. Christian Ott, who supported me and introduced me to the exciting field of stability theory in robotics. Moreover, I wish to thank Daniel Leidner and Dr. Thomas Wimböck, with whom I collaborated in a very productive way resulting in several valuable publications in the field of whole-body control.

Without the excellent and continuous maintenance of the robot software and hardware by Florian Schmidt, Robert Burger, and Werner Friedl, the numerous experiments in this book would not have been possible at all. Moreover, I would like to thank my colleagues Jens Reinecke, Dr. Maxime Chalon, Dr. Florian Petit, Dominic Lakatos, Andreas Stemmer, Philipp Stratmann, and my former students Melanie Kimmel and Kristin Bussmann for the fruitful discussions and their support.

My gratitude also goes to Prof. Gerd Hirzinger, who gave me the opportunity to work at the DLR and use the remarkable robotic systems for my research. Furthermore, I would like to thank Dr. Paul Kotyczka and Prof. Boris Lohmann for the great cooperation between DLR and TUM.

I would like to thank Prof. Bruno Siciliano, the team of Springer, and the committee of the Georges Giralt Ph.D. Award for giving me the opportunity to publish my research findings in the Springer Tracts in Advanced Robotics.

Thanks and love to my parents Christine and Rainer, and my sister Kerstin, who have always encouraged and supported me. Finally, I thank my beloved wife Ann-Kristin. Without her patience and love, this work would have never been completed.

Munich Alexander Dietrich
April 2016

Contents

1 Introduction 1
1.1 Motivation 1
1.2 Related Work 3
1.3 Problem Statement 5
1.4 Concept of Whole-Body Impedance 6
1.5 Contributions and Overview 8

2 Fundamentals 13
2.1 Robot Kinematics and Dynamics 13
2.1.1 Forward Kinematics, Jacobian Matrices, and Power Ports 13
2.1.2 Derivation of the Equations of Motion 14
2.1.3 Rigid Body Dynamics 16
2.2 Compliant Motion Control of Robotic Systems 16
2.2.1 Impedance Control 17
2.2.2 Admittance Control 18
2.3 Humanoid Robot Rollin' Justin 19
2.3.1 Design and Hardware 19
2.3.2 Modeling Assumptions 21

3 Control Tasks Based on Artificial Potential Fields 23
3.1 Self-Collision Avoidance 24
3.1.1 Geometric Collision Model 25
3.1.2 Repulsive Potential 26
3.1.3 Damping Design 28
3.1.4 Control Design 31
3.1.5 Experiments 33
3.2 Singularity Avoidance for Nonholonomic, Wheeled Platforms 34
3.2.1 Instantaneous Center of Rotation 34
3.2.2 Controllability and Repulsion 36
3.2.3 Effect on the Instantaneous Center of Rotation 37
3.2.4 Effect on the Wheel 39

3.2.5 Control Design . . . 39
3.2.6 Simulations and Experiments . . . 40
3.3 Posture Control for Kinematically Coupled Torso Structures . . . 44
3.3.1 Model of the Torso of Rollin' Justin . . . 45
3.3.2 Kinematic Constraints . . . 45
3.3.3 Dynamic Constraints . . . 46
3.3.4 Control Design . . . 49
3.3.5 Experiments . . . 49
3.4 Classical Objectives in Reactive Control . . . 50
3.4.1 Cartesian Impedance . . . 50
3.4.2 Manipulator Singularity Avoidance . . . 51
3.4.3 Avoidance of Mechanical End Stops . . . 52
3.5 Summary . . . 52

4 Redundancy Resolution by Null Space Projections . . . 55
4.1 Strictness of the Hierarchy . . . 56
4.1.1 Successive Projections . . . 56
4.1.2 Augmented Projections . . . 57
4.2 Consistency of the Projections . . . 58
4.2.1 Static Consistency . . . 59
4.2.2 Dynamic Consistency . . . 60
4.2.3 Stiffness Consistency . . . 64
4.3 Comparison of Null Space Projectors . . . 65
4.3.1 Simulations . . . 66
4.3.2 Experiments . . . 70
4.3.3 Discussion . . . 75
4.4 Unilateral Constraints in the Task Hierarchy . . . 78
4.4.1 Basics . . . 79
4.4.2 Ensuring Continuity . . . 81
4.4.3 Simulations . . . 86
4.4.4 Experiments . . . 87
4.4.5 Discussion . . . 95
4.5 Summary . . . 97

5 Stability Analysis . . . 99
5.1 Whole-Body Impedance with Kinematically Controlled Platform . . . 99
5.1.1 Subsystems . . . 100
5.1.2 Control Design . . . 106
5.1.3 Proof of Stability . . . 107
5.1.4 Experiments . . . 109
5.1.5 Discussion . . . 114
5.2 Multi-Objective Compliance Control . . . 116
5.2.1 Problem Formulation . . . 118
5.2.2 Hierarchical Dynamics Representation . . . 120

5.2.3 Control Design 126
5.2.4 Proof of Stability 128
5.2.5 Simulations and Experiments 131
5.2.6 Discussion 137
5.3 Summary 139

6 Whole-Body Coordination 141
6.1 Order of Tasks in the Hierarchy 142
6.2 Implementation on Rollin' Justin 144
6.3 Summary 148

7 Integration of the Whole-Body Controller into a Higher-Level Framework 151
7.1 Intelligent Parameterization of the Whole-Body Controller 153
7.2 Communication Channel Between Controller and Planner 153
7.3 Real-World Applications for a Service Robot 154
7.4 Summary 156

8 Summary 157

Appendix A: Workspace of the Torso of Rollin' Justin 161

Appendix B: Null Space Definitions and Proofs 163

Appendix C: Proofs for the Stability Analysis 169

Appendix D: Stability Definitions 175

References 177

Symbols and Abbreviations

In this monograph, all scalar quantities are described by plain letters (e.g., n, λ, c_1). Matrices and vectors are printed in bold (e.g., $\boldsymbol{x}$, $\boldsymbol{g}(\boldsymbol{q})$, $\boldsymbol{M}(\boldsymbol{q})$). Total derivatives with respect to time t are abbreviated by dots (e.g., $\dot{\boldsymbol{x}} = \frac{d}{dt}\boldsymbol{x}$, $\ddot{\boldsymbol{x}} = \frac{d^2}{dt^2}\boldsymbol{x}$).

Several variables in the following list appear with different subscripts, superscripts, additional symbols, and various dimensions in this book. Here, the quantities are listed and generally described without further specification. The specific meaning becomes clear when the respective variable is introduced in the text. Note that this list of variables is not complete, but it only contains quantities which appear at several places in the book or are of prominent importance.

Symbols

$\mathcal{A}$	Set
c	Constant
d	Distance
i, j	Indices (for numbering)
l	Kinematic parameter
m	Number or mass
n	Number (e.g., joints, collision pairs)
S	Storage function
t	Time
V	Potential function
λ	Leg length of wheeled platform or damping parameter
σ	Singular value
ξ	Damping ratio
$\boldsymbol{C}$	Coriolis/centrifugal matrix
$\boldsymbol{D}$	Damping matrix
$\boldsymbol{f}$	Vector function describing a task with task coordinates $\boldsymbol{x}$

$\boldsymbol{F}$	Vector of (generalized) Cartesian forces
$\boldsymbol{g}$	Vector of gravity torques
$\boldsymbol{J}$	Jacobian matrix
$\boldsymbol{K}$	Stiffness matrix
$\boldsymbol{M}$	Inertia matrix
$\boldsymbol{N}$	Null space projection matrix
$\boldsymbol{p}$	Vector describing a point in space
$\boldsymbol{q}$	Vector of link-side joint angles
$\boldsymbol{r}$	Vector of mobile base coordinates (in motion direction)
$\boldsymbol{U}$, $\boldsymbol{S}$, $\boldsymbol{V}$	Matrix components in a singular value decomposition
$\boldsymbol{v}$	Vector of local null space velocities
$\boldsymbol{w}$	Vector of wheel positions (steering and propulsion)
$\boldsymbol{W}$	Weighting matrix
$\boldsymbol{x}$	Vector of Cartesian coordinates
$\boldsymbol{X}$	Submatrix of $\boldsymbol{V}$ (range space of Jacobian matrix $\boldsymbol{J}$)
$\boldsymbol{y}$	Vector describing the robot configuration (with mobile base)
$\boldsymbol{Y}$	Submatrix of $\boldsymbol{V}$ (null space of Jacobian matrix $\boldsymbol{J}$)
$\boldsymbol{Z}$	Matrix describing the null space base of a Jacobian matrix $\boldsymbol{J}$
$\boldsymbol{\Lambda}$	Reflected inertia matrix (e.g., in Cartesian directions $\boldsymbol{x}$)
$\boldsymbol{\mu}$	Coriolis/centrifugal matrix in hierarchical dynamics representation
$\boldsymbol{\tau}$	Vector of joint torques

Abbreviations

adm	Admittance
aug	Augmented
cmd	Command
des	Desired
e.g.	*Exempli gratia* (for example)
ext	External
i.e.	*Id est* (that is)
icr	Instantaneous center of rotation
kin	Kinematic
max	Maximum
mes	Mechanical end stop(s)
msa	Manipulator singularity avoidance
min	Minimum
ref	Reference
sca	Self-collision avoidance
sap	Singularity avoidance for platform
suc	Successive

tws	Torso workspace
vir	Virtual
w.r.t.	With respect to
DOF	Degree(s) of freedom
DLR	Deutsches Zentrum für Luft- und Raumfahrt e.V. (German Aerospace Center)
OSF	Operational space formulation
TUM	Technische Universität München (Technical University of Munich)
TCP	Tool center point

Chapter 1
Introduction

The robotics research in the recent past has led to the development of an increasing number of mobile humanoid robots. They can be employed in a great diversity of applications such as service robotics, the cooperation with humans in industry, or the autonomous operation in hazardous areas where humans would be in danger. All of these use cases involve dynamic, unpredictable, and partially unstructured environments, where physical contacts are inevitable and actually necessary for the task completion. The high requirements on the humanoid robots urge the designers to develop suitable whole-body control techniques in order to properly operate the systems.

The first chapter introduces the reader to the overall topic of this book. After the motivation in Sect. 1.1, the state of the art in whole-body control is recapitulated in Sect. 1.2. Following this, the problem is stated (Sect. 1.3) and the approach is sketched (Sect. 1.4). Chapter 1 ends with the contribution of this monograph and the overview of the content (Sect. 1.5).

1.1 Motivation

The idea of humanoid robots has inspired researchers, artists, and authors for a long time. The most popular novelist is Isaac Asimov who wrote futuristic science fiction literature about the coexistence of robots and humans. In his story *Runaround* from 1942, Asimov introduced the famous *Three Laws of Robotics*:

1. "*A robot may not injure a human being or, through inaction, allow a human being come to harm.*"
2. "*A robot must obey the orders given to it by human beings, except where such orders would conflict with the First Law.*"

A. Dietrich, *Whole-Body Impedance Control of Wheeled Humanoid Robots*,
Springer Tracts in Advanced Robotics 116, DOI 10.1007/978-3-319-40557-5_1

3. "*A robot must protect its own existence as long as such protection does not conflict with the First or Second Law.*"

Although the above laws originate from science fiction, rules of that kind are also concluded by engineers during the process of developing humanoid robots. In other words, the core requirements from a developer's perspective are also expressed by Asimov's Laws: *safety* is a key issue; protecting the human is more important than self-protection of the robot; the main purpose of the robot is to *obey and perform tasks* to support the human; *priorities among the objectives* are essential; proper operation is only possible if the *hierarchy* is strictly satisfied. All these conclusions will play an essential role throughout this book.

The tremendous technological progress in the last decades has made it possible to build the first humanoid robots. Famous legged systems are ASIMO [SWA + 02] and its successors, HRP-4 [KKM + 11], or LOLA [LBU09], to mention just a few. Another class of humanoid robots is characterized by anthropomorphic upper bodies mounted on wheeled mobile platforms: ARMAR-III [ARS + 06], PR2 [BRJ + 11], TWENDY-ONE [IS09], Robonaut 2 [DMA + 11], or Rollin' Justin [BWS + 09]. These systems are predestined to be employed in human environments such as households. Legged humanoid robots are potentially more versatile than their wheeled counterparts when considering mobility in domestic environments with stairs or doorsteps, for example. Yet, most complex service tasks have been executed by wheeled robots only. The advantage of a wheeled mobile platform is to focus on sophisticated manipulation skills without the necessity of balancing and stabilizing the gait. Therefore, one may expect wheeled robots to occupy an important place in the future of service robotics. A start has already been made by the first commercially available robotic vacuum cleaners and lawn mowers.

The versatility and dexterity of the human body can be partially attributed to its large number of actuated degrees of freedom (DOF). Envision a service task such as cleaning a window with a wiper. To execute it, one needs at least six DOF because the wiper trajectory requires a complete definition in space (position and orientation). The kinematic redundancy w.r.t. this task allows to simultaneously accomplish further goals, for example avoiding collisions with the wall, optimizing the posture to minimize the stress on the spinal discs, or looking around to observe the environment. All these subtasks can be performed without suspending the main cleaning task by using the redundant DOF. This monograph will present a *whole-body controller* for humanoid robots with appropriate redundancy resolution by considering the structure in its entirety instead of treating the subsystems (e.g. arms, torso, locomotion system) separately.

As mentioned above, safety is a key issue. Especially when considering households or crowded places, the environment is often unknown, unstructured, dynamic, and unpredictable due to the presence of human beings. The requirements clearly differ from classical industrial applications in factories where robots are usually caged and physically separated from the user. In human environments the robot has to feature a certain degree of compliance to be able to instantaneously react in case of contacts and physical interactions. While this can be achieved by passive

Fig. 1.1 Wheeled humanoid lightweight robot Rollin' Justin developed at the Institute of Robotics and Mechatronics at the German Aerospace Center (DLR)

elements such as mechanical springs, the concept of active compliance is dominating the field. The most popular method is impedance control [Hog85], where a desired mass-spring-damper behavior is specified. The implementation of an impedance requires commanding forces or torques, respectively. One possible realization is to utilize joint torque sensing to implement torque control like in Rollin' Justin [ASOH07], see Fig. 1.1. Compared to many classical techniques in control theory such as pole placement [FW67] or backstepping [KKK95], an impedance controller is easy and intuitive to parameterize because the design parameters (stiffness, inertia) have a direct relation to the physical world. Furthermore, active impedance control is known to be very robust, which is an important aspect for robots in dynamic, unpredictable, and unknown environments. The methods have already found their way into industrial applications [ASEG+08] where compliance and robustness are needed such as in assembly tasks. Their success in physical interaction scenarios predestines them for the next evolutionary step, that is, to enter the field of service robotics and the installation in households and domestic environments.

1.2 Related Work

In the 1980s, an innovative idea revolutionized robot motion planning and control: the task space or operational space concept [Kha87]. Instead of defining objectives in the abstract joint space of the robot, task coordinates with intuitive interpretation were introduced. The most common example is certainly the Cartesian space of the end-effector of a manipulator. This simplification inevitably led to the question of an adequate redundancy resolution since most robotic systems have more actuated DOF than the dimension of the task space. The invention of the so-called null space projection [BHB84, NHY87, HS87] partially enabled to deal with this issue and resolve the redundancy by performing additional tasks in the null space of the end-effector task, that is, without disturbing the end-effector task execution. Since then, a wide range of sophisticated frameworks have been developed based on these essential techniques and have extended them by useful improvements in computational aspects [BB98], singularity-robustness [NH86], motion generation [BK02], or the extension to more complex multi-priority hierarchies [SS91], for example.

In the context of whole-body control, attractive and repulsive artificial potential fields [Kha86] are the most frequently employed robot control methods to accomplish tasks. Numerous solutions to specific control problems exist such as in the fields of collision avoidance [SGJG10, DSASO+07], manipulator singularity avoidance [Ott08], singularity avoidance for wheeled platforms [CPHV09], dual-arm manipulation [WOH07], or joint limit avoidance [MCR96]. Particularly since the early 2000s, more and more whole-body controllers that implement several of these objectives simultaneously have been released thanks to the availability of suitable simulation models and hardware. Seminal results in multi-task hierarchies were obtained by Sentis et al. [SK05, SPK10] and Khatib et al. [KSPW04, KSP08] in concept and simulation. Hammam et al. [HOD10] simulated such a framework on ASIMO, Sadeghian et al. [SVKS13] applied a multi-priority controller to the model of a torque-controlled lightweight robot.

The number of implementations on real robots is also steadily increasing. Nevertheless, one will only find a small number in the literature so far. Yoshikawa and Khatib [YK09] implemented an inverse model to transform commanded torques into velocity signals so that soft contact behavior can be realized on a position-controlled robot. An open-source software package for whole-body compliance has been released by Philippsen et al. [PSK11] and has already been utilized on the robots PR2 and Dreamer. An attractor-based multi-task method has been implemented on the legged COMAN robot by Moro et al. [MGG+13]. Nagasaka et al. implemented a whole-body algorithm for diverse motion objectives on a 21-DOF wheeled system [NKS+10], but several crucial aspects have not been considered such as singularity treatment or collision avoidance. A whole-body compliance controller for kinesthetic teaching of torque-controlled robots has been recently proposed by Ott et al. [OHL13]. The implementation was carried out on TORO, a legged humanoid based on the DLR lightweight robot technology which is also used in Rollin' Justin.

This list of whole-body control concepts could be easily continued. So what is the point of further research in the field and the justification for another approach? The following section will reveal fundamental lacks of the existing techniques, which hinder their wide use and their employment in commercial service robotics so far.

1.3 Problem Statement

As mentioned above, humanoid robots are available for a few years. However, if one demands a joint torque interface for compliant physical interaction, the quantity of readily available systems reduces to a small number. Consequently, only little research and experimental validation on these systems have been done so far. Although new developments from research in humanoid robotics appear more and more frequently in the mass media, for example in movies, television, or newspapers, it is actually only the beginning of a long-term process that involves iterations between theoretical advances, improved hardware, and the experimental validation.

A central issue of most whole-body control concepts is the lack of valid stability analyses and proofs. The overwhelming majority of approaches is based on the classical *Operational Space Formulation* (OSF) [Kha87], which leads to the dynamic decoupling of the prioritized tasks through feedback linearization. However, the OSF does not provide overall stability. Only exponential stability on the main task level (highest priority) can be shown so far. Nakanishi et al. [NCM+08] compared eight of the most commonly used OSF controllers from a theoretical and empirical perspective and concluded that "*null space dynamics so far resist insightful general analytical investigations (...) If stability could be proven for this family of operational space controllers, operational space control would be lifted to a more solid foundation*". The problem is related to the dynamics of all subordinated (null space) tasks. The authors state that "*the exact behavior of the null space dynamics cannot be determined easily (...) This difficulty of understanding the null space stability properties is, however, a problem that is shared by all operational space controllers. So far, only empirical evaluations can help to assess the null space robustness*". In safety-critical environments such as households or crowded places, the application of techniques without knowledge of the overall stability properties can pose a significant obstacle. Of course, one might argue that even without a theoretical stability analysis, many methods in robot control have proven successful. But a lack of proof of stability can have severe, practical consequences concerning stability and robustness, even if only under particular circumstances or in extreme situations. In Chap. 5, such an example is given where the robot operates properly in most cases, but a slight change in the impedance controller parameters suddenly destabilizes the system. The subsequent, formal stability analysis identifies the reasons, solves the problem by an additional control action and guarantees stability for all controller parameterizations. As Kurt Lewin (1890–1947), one of the pioneers of modern psychology, once said, "*There is nothing so practical as a good theory*". This statement is also valid in robotics, especially when the consequences affect human lives: Who would actually buy a robot for his own home, if the manufacturer is not able to guarantee safe and reliable operation in *any* case?

Another problem with whole-body controllers based on the OSF concerns the external forces and torques. Either they are not considered in the control law and the system does not feature a specified compliant behavior w. r. t. its environment in consequence, or compliance has to be paid for by their measurement/estimation and feedback. While external loads applied at the end-effector can be measured if a force-torque sensor is mounted at the tip, external forces and torques exerted on other parts of the robot structure cannot be easily identified.

Whole-body control is a complex challenge and most existing frameworks cover subdomains only. A proper approach should fulfill a large number of requirements simultaneously:

- Many different objectives must be considered such as safety features, task execution, optimization criteria, and so on.
- The order of priority has to be satisfied and the task hierarchy must be flexible to manage unilateral constraints, dynamic modifications, and singularities.
- Stability of the closed loop must be ensured.
- The robot has to physically interact with its environment in a compliant way so that no human in its workspace may come to harm.
- The approach must be validated experimentally.

A solution for all of these issues in one unified framework is desirable.

Once a control framework is available, it is possible to merge it with a higher-level artificial intelligence (AI) instance. State-of-the-art whole-body concepts do not yet consider the integration of the controller into such a higher-level framework. The key is to establish a two-way communication channel between the controller and the AI module. Then the AI can properly parameterize the controller, choose reasonable control objectives, determine application-specific task hierarchies, and command the respective trajectories. On the other hand, the whole-body controller (hard real-time) can feedback useful information about the task execution so that the (non-deterministic) AI instance can replan in order to find global solutions, e.g. when the controller is stuck in a local minimum due to altered environmental conditions.

1.4 Concept of Whole-Body Impedance

The title of this monograph is "Whole-Body Impedance Control of Wheeled Humanoid Robots". While the term "wheeled humanoid robot" is self-explanatory, the term "whole-body impedance control" may be less obvious. In the course of this work, various impedance control tasks are established in a hierarchical order. A representative setup is an operational space impedance controller which is applied to the complete, multi-DOF body of a humanoid robot as illustrated in Fig. 1.2. A desired, compliant contact behavior of the end-effector is assigned in its operational space, e.g. the six-dimensional Cartesian space of the hand. The impedance can be intuitively interpreted as a mass-spring-damper system with actively controlled *virtual* spring and *virtual* damper.[1] Although these elements are only virtually applied, the

[1] One may also shape the perceived inertia, but the case of active spring and damper control (*compliance control*) is emphasized in this book because it does not require the problematic feedback of external forces and torques. Moreover, the interaction behavior is more natural when keeping the natural inertia.

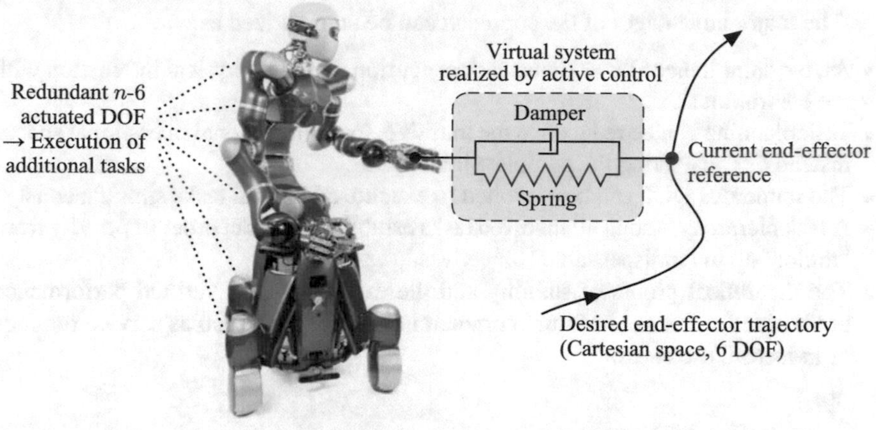

Fig. 1.2 Example of whole-body impedance control illustrated on a wheeled humanoid robot with n actuated DOF

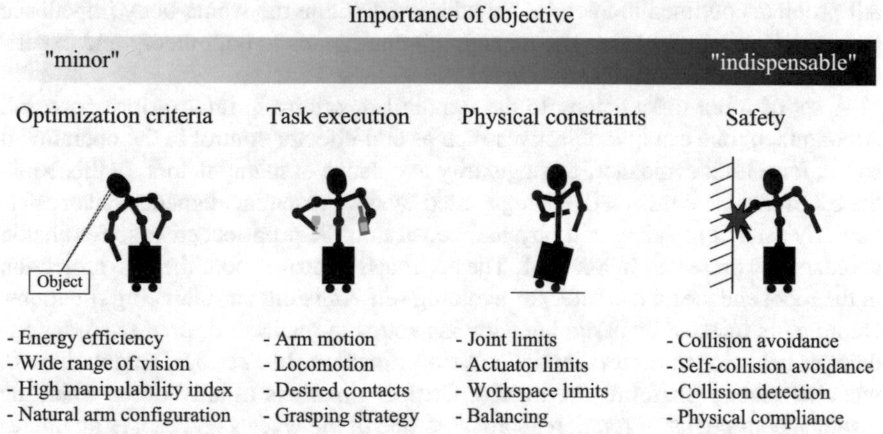

Fig. 1.3 A wide variety of possible objectives that can be assigned to the robot within the whole-body impedance framework

human will actually feel the specified physical compliance while interacting with the robot. The approach enables complex task execution by planning in low-dimensional, intuitive spaces instead of considering the complete, bulky configuration space. The remaining DOF, which can be controlled without affecting the operational space impedance, can be utilized to accomplish additional tasks such as collision avoidance or the optimization of various criteria. Figure 1.3 shows a wide range of possible objectives. Since there are usually "more important" and "less important" goals, the whole-body impedance framework is developed in a way such that a desired control task hierarchy can be realized.

The major advantages of the approach can be summarized as:

- Active compliance allows robust task execution and safe physical interaction with the environment.
- Task planning can be reduced to the intuitive, low-dimensional operational spaces instead of considering the complete joint space.
- The numerous DOF can be exploited to execute additional tasks simultaneously.
- A task hierarchy among all involved tasks establishes a strict order of priority from "minor" up to "indispensable" objectives.
- The theoretical proofs of stability and the experimentally verified performance predestine for the use in future commercial applications such as service robotics or industrial use cases.

1.5 Contributions and Overview

All problems outlined in Sect. 1.3 are addressed within the whole-body impedance framework developed here. The monograph contributes to both theory and experimental validation in this research field.

A set of robot tasks belong to the standard repertoire in multi-priority control. Among them one can find objectives such as end-effector control in the operational space, joint limit avoidance, or singularity avoidance of manipulators. In this book, the set of available tasks will be augmented by new, essential elements that are particularly related to safety and physical constraints: A proprioceptive self-collision avoidance is presented in Sect. 3.1. The approach improves both the self-protection of the robot and the human safety by avoiding self-collisions and clamping situations. Damping is injected by assigning damping ratios to realize a desired mass-spring-damper behavior in each potential collision direction. In Sect. 3.2, singularities in wheeled mobile platforms are treated. Critical situations usually occur when the instantaneous center of rotation approaches one of the wheels and causes an unpredictable and dangerous platform behavior. The proposed method reactively avoids these critical configurations. Section 3.3 addresses workspace limitations of tendon-coupled torso structures such as the one in Rollin' Justin. The constraints due to the coupling restrict both the kinematic and the dynamic workspace of the robot. A repulsion from these boundaries is proposed. The three reactive control methods in Chap. 3 are experimentally validated on Rollin' Justin.

The concept of null space projections is addressed in Chap. 4. Beside a survey of all relevant implementations in torque control, the popular dynamically consistent redundancy resolution [Kha87] is generalized. Moreover, a new type of null space projection is introduced, namely the stiffness-consistent approach. This new method allows to exploit the mechanical springs in parallel elastic actuators (PEA) within the task hierarchy. A further contribution in Chap. 4 is the incorporation of unilateral constraints. That makes it possible to deal with singular Jacobian matrices and allows

Table 1.1 Main publications on which this monograph is based

Reference	Description
Journal: [DWASH12b]	A. Dietrich, T. Wimböck, A. Albu-Schäffer, and G. Hirzinger. Reactive Whole-Body Control: Dynamic Mobile Manipulation Using a Large Number of Actuated Degrees of Freedom. *IEEE Robotics & Automation Magazine*, 19(2):20–33, June 2012
Journal: [DWASH12a]	A. Dietrich, T. Wimböck, A. Albu-Schäffer, and G. Hirzinger. Integration of Reactive, Torque-Based Self-Collision Avoidance Into a Task Hierarchy. *IEEE Transactions on Robotics*, 28(6):1278–1293, December 2012
Journal: [DOAS15]	A. Dietrich, C. Ott, and A. Albu-Schäffer. An overview of null space projections for redundant, torque-controlled robots. *International Journal of Robotics Research*, 34(11):1385–1400, September 2015
Journal: [ODAS15]	C. Ott, A. Dietrich, and A. Albu-Schäffer. Prioritized Multi-Task Compliance Control of Redundant Manipulators. *Automatica*, 53:416–423, March 2015.
Journal: [DBP+16]	A. Dietrich, K. Bussmann, F. Petit, P. Kotyczka, C. Ott, B. Lohmann, and A. Albu-Schäffer. Whole-body impedance control of wheeled mobile manipulators: Stability analysis and experiments on the humanoid robot Rollin' Justin. *Autonomous Robots*, 40(3):505–517, March 2016
Journal: [LDBAS16]	D. Leidner, A. Dietrich, M. Beetz, and A. Albu-Schäffer. Knowledge-enabled parameterization of whole-body control strategies for compliant service robots. *Autonomous Robots*, 40(3):519–536, March 2016
Conference: [DWASH11]	A. Dietrich, T. Wimböck, A. Albu-Schäffer, and G. Hirzinger. Singularity Avoidance for Nonholonomic, Omnidirectional Wheeled Mobile Platforms with Variable Footprint. In *Proc. of the 2011 IEEE International Conference on Robotics and Automation*, pages 6136–6142, May 2011
Conference: [DWT+11]	A. Dietrich, T. Wimböck, H. Täubig, A. Albu-Schäffer, and G. Hirzinger. Extensions to Reactive Self-Collision Avoidance for Torque and Position Controlled Humanoids. In *Proc. of the 2011 IEEE International Conference on Robotics and Automation*, pages 3455–3462, May 2011
Conference: [DWAS11]	A. Dietrich, T. Wimböck, and A. Albu-Schäffer. Dynamic Whole-Body Mobile Manipulation with a Torque Controlled Humanoid Robot via Impedance Control Laws. In *Proc. of the 2011 IEEE/RSJ International Conference on Intelligent Robots and Systems*, pages 3199–3206, September 2011
Conference: [DASH12]	A. Dietrich, A. Albu-Schäffer, and G. Hirzinger. On Continuous Null Space Projections for Torque-Based, Hierarchical, Multi-Objective Manipulation. In *Proc. of the 2012 IEEE International Conference on Robotics and Automation*, pages 2978–2985, May 2012
Conference: [DOAS13]	A. Dietrich, C. Ott, and A. Albu-Schäffer. Multi-Objective Compliance Control of Redundant Manipulators: Hierarchy, Control, and Stability. In *Proc. of the 2013 IEEE/RSJ International Conference on Intelligent Robots and Systems*, pages 3043–3050, November 2013
Conference: [DKW+14]	A. Dietrich, M. Kimmel, T. Wimböck, S. Hirche, and A. Albu-Schäffer. Workspace Analysis for a Kinematically Coupled Torso of a Torque Controlled Humanoid Robot. In *Proc. of the 2014 IEEE International Conference on Robotics and Automation*, pages 3439–3445, June 2014
Conference: [LDS+14]	D. Leidner, A. Dietrich, F. Schmidt, C. Borst, and A. Albu-Schäffer. Object-Centered Hybrid Reasoning for Whole-Body Mobile Manipulation. In *Proc. of the 2014 IEEE International Conference on Robotics and Automation*, pages 1828–1835, June 2014
Patent: [DBOAS14]	A. Dietrich, K. Bussmann, C. Ott, and A. Albu-Schäffer. Ganzkörperimpedanz für mobile Roboter. German patent application No. 10 2014 226 936, patented on December 23, 2014

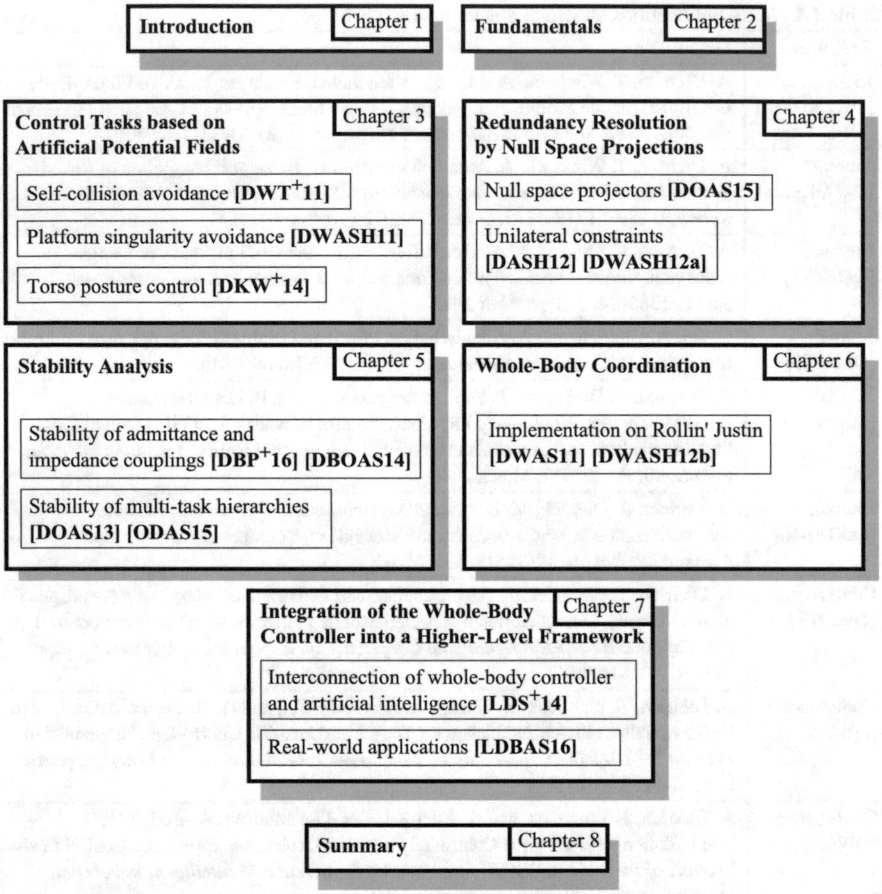

Fig. 1.4 Graphical overview of the chapters, the main topics, and the relation to the publications

flexible and dynamic hierarchies. All proposed concepts are validated on the real robot.

The stability analysis is treated in Chap. 5. The first contribution is the analysis of a generic humanoid robot with wheeled platform and torque-controlled upper body in Sect. 5.1. Since a mobile base is normally kinematically controlled due to the rolling constraints, the integration into an overall impedance control framework is not straightforward. Via admittance couplings this issue is solved and by smart adaptation of the control law, asymptotic stability of the equilibrium and passivity properties are shown. The second contribution in Sect. 5.2 is a proof of stability for arbitrarily complex multi-task hierarchies, which can also be applied to legged systems or other classes of robots. A novel, priority-based dynamics representation is derived such that asymptotic stability can be concluded due to the beneficial properties of the formulation of the equations of motion and the particular control law. As mentioned in Sect. 1.3 and by Nakanishi et al. [NCM+08], such a feature has not been provided elsewhere yet. Again, all results are experimentally validated on Rollin' Justin. The stability analyses in Chap. 5 can be combined such that the proofs are universally valid for torque-controlled robots with wheeled mobile platforms and a stack of hierarchically arranged whole-body control tasks.

The contributions in Chap. 6 are mainly of experimental nature. The whole-body controller is implemented on Rollin' Justin involving the theoretical results of this monograph. A detailed analysis is conducted and the performance of the proposed approach is confirmed.

Chapter 7 contributes to the combination of local methods (control) and global methods (planning). This is actually a new field of research that has not been treated thoroughly in the robotics community so far. In this respect, Chap. 7 represents the first steps on the way towards the integration of whole-body control into higher-level frameworks involving artificial intelligence, so that the ideal of an autonomous and intelligent humanoid robot will be realized one day.

The research findings reported in this book resulted in six journal articles, seven conference papers on the main robotics congresses, and a patent that is currently under review. These main publications on which the monograph is based are summarized in Table 1.1 and related to the respective chapters of this book in Fig. 1.4. Furthermore, two journal articles [PDAS15, CDG14] and six conference papers [BSW+11, BBW+11, LGP+13, RDC14, LGDAS14, ODR14] have been co-authored, which are related to the topic but not integrated in the monograph.

Chapter 2
Fundamentals

This chapter briefly reviews the basics that are required for the theoretical investigations and the practical implementations in this monograph. That comprises fundamentals in kinematics and dynamics (Sect. 2.1), active control for compliant interaction behavior (Sect. 2.2), and details on the hardware and modeling assumption on the wheeled humanoid robot Rollin' Justin, which has been chosen as the platform for the experimental validations (Sect. 2.3).

2.1 Robot Kinematics and Dynamics

The following sections introduce some basic kinematic and dynamic matters of special importance in the context of this work. For a complete and more detailed version the reader is referred to the standard literature [Pau83, Cra89, Yos90, MLS94, KD02, SK08].

Most robotic designs are based on revolute joints rather than prismatic joints. Thus, one has to deal with torques instead of forces on joint level. In this book the term *joint torque* is mostly used, but the extension to generalized joint forces (including forces and torques) can be made. By default, external loads are referred to as *external forces* since physical contact usually occurs from contact surface to contact surface, hence a force is more common. However, the extension to generalized external forces (including forces and torques) can be made without loss of generality. The simplified notations are used for the sake of brevity.

2.1.1 *Forward Kinematics, Jacobian Matrices, and Power Ports*

A typical robotic system is described by $\boldsymbol{q} \in \mathbb{R}^n$ joint coordinates, where n is the number of degrees of freedom (DOF). The operational space, e.g. the

A. Dietrich, *Whole-Body Impedance Control of Wheeled Humanoid Robots*,
Springer Tracts in Advanced Robotics 116, DOI 10.1007/978-3-319-40557-5_2

workspace of the end-effector, is usually described by $m \leq n$ coordinates denoted $\boldsymbol{x}(\boldsymbol{q}, \mathcal{L}) \in \mathbb{R}^m$. The kinematic parameterization $\mathcal{L}$ is usually constant and disregarded in the notations. The robot is assumed to be rigid, the only relative motion is along/about the n joint axes.

The most common representation of $\boldsymbol{x}(\boldsymbol{q})$ is in Cartesian coordinates. If $m = n$, then the robot is called non-redundant w. r. t. the operational space with dimension m. If $m < n$, then the manipulator is kinematically redundant and can execute additional tasks by utilizing the $(n-m)$-dimensional *null space* while not disturbing the operational space task.[1] The forward kinematics of the considered robot are described by the mapping $\boldsymbol{q} \mapsto \boldsymbol{x}$, while the inverse kinematics $\boldsymbol{x} \mapsto \boldsymbol{q}$ in a redundant robot is ambiguous and requires additional constraints to be resolved.

The differential and velocity relation between the joint space and the operational space

$$\Delta\boldsymbol{x}(\boldsymbol{q}) = \frac{\partial \boldsymbol{x}(\boldsymbol{q})}{\partial \boldsymbol{q}}\Delta\boldsymbol{q} = \boldsymbol{J}(\boldsymbol{q})\Delta\boldsymbol{q}, \tag{2.1}$$

$$\dot{\boldsymbol{x}}(\boldsymbol{q}, \dot{\boldsymbol{q}}) = \boldsymbol{J}(\boldsymbol{q})\dot{\boldsymbol{q}}, \tag{2.2}$$

necessitates a further quantity $\boldsymbol{J}(\boldsymbol{q}) \in \mathbb{R}^{m \times n}$, the Jacobian matrix. The total time derivative $d\{\}/dt$ of a function $\{\}$ is abbreviated as $\dot{\{\}}$ in this book. Based on the geometric point of view (2.1) and (2.2), one can find a simple relation between operational space forces $\boldsymbol{F} \in \mathbb{R}^m$ and joint torques $\boldsymbol{\tau} \in \mathbb{R}^n$:[2]

$$\boldsymbol{\tau} = \boldsymbol{J}(\boldsymbol{q})^T \boldsymbol{F} \tag{2.3}$$

The application of joint torques with (2.3) is called a *Jacobian transposed* approach. This concept is adopted here and constitutes a basic prerequisite for the methods developed in the later chapters. The variables in (2.2) and (2.3) describe either a flow ($\dot{\boldsymbol{q}}$ or $\dot{\boldsymbol{x}}$) or an effort ($\boldsymbol{\tau}$ or $\boldsymbol{F}$). The associated terms build *power ports*, since they define a power through $\dot{\boldsymbol{q}}^T\boldsymbol{\tau}$ and $\dot{\boldsymbol{x}}^T\boldsymbol{F}$, respectively. Via such a port, the system can exchange energy with its environment. All relations are illustrated in Fig. 2.1.

2.1.2 Derivation of the Equations of Motion

Two basic formalisms are briefly reviewed that yield the dynamic equations of a robot. These sections explain the derivation in a nutshell only. A more detailed version can be found in the standard literature listed in Sect. 2.1.

---

[1] Motions in the null space are also denoted *internal motions*.

[2] Note that $\boldsymbol{F}$ may also contain torques, e.g. in the case of a full operational space wrench $\boldsymbol{F} \in \mathbb{R}^6$ with three forces and three torques. Furthermore, $\boldsymbol{\tau}$ may also contain force elements in case of prismatic joints.

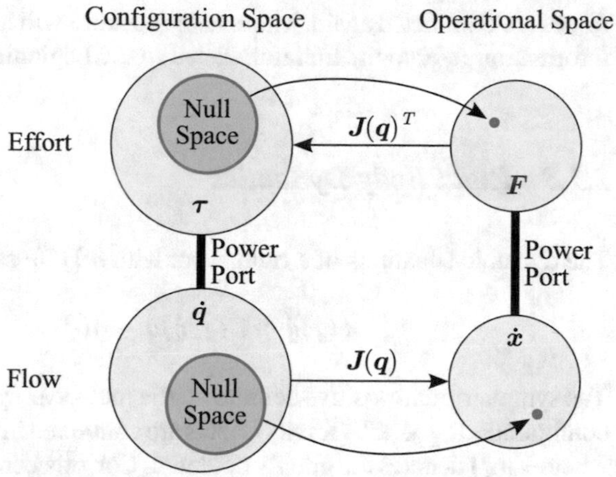

Fig. 2.1 Relations between $\dot{x}$, F, $\dot{q}$, and τ for a redundant manipulator with $m < n$

2.1.2.1 Lagrange Formalism

The Lagrange formalism is an energy-based technique to obtain the dynamic equations. An n-DOF system with joint values $q \in \mathbb{R}^n$ and joint velocities $\dot{q} \in \mathbb{R}^n$ is described by the so-called Lagrangian

$$L(q, \dot{q}) = T(q, \dot{q}) - V(q), \tag{2.4}$$

which is given by the kinetic energy $T(q, \dot{q})$ of the system minus its potential energy $V(q)$. Then the dynamic equations can be obtained by evaluating

$$\frac{d}{dt}\left(\frac{\partial L(q, \dot{q})}{\partial \dot{q}}\right)^T - \left(\frac{\partial L}{\partial q}\right)^T = Q. \tag{2.5}$$

The vector $Q \in \mathbb{R}^n$ contains the generalized joint forces $\tau \in \mathbb{R}^n$, the external loads $\tau_{\text{ext}} \in \mathbb{R}^n$, and non-conservative generalized forces such as friction. The main advantage of the method is the simple analytical determination of the kinetic and potential energy. But the computational burden of the method makes it unsuitable for large systems with many DOF. The computational effort for an n-link robot is of order $\mathcal{O}(n^4)$ while it is only $\mathcal{O}(n)$ with the iterative Newton–Euler formalism. See [Yos90] for a more detailed comparison.

2.1.2.2 Iterative Newton–Euler Formalism

The iterative Newton–Euler algorithm requires the evaluation of Euler's first and second law for each link of the robot. All constraining forces have to be calculated explicitly. Finally, the equations of all links are combined and the constraining forces are eliminated again. The exact procedure will not be detailed here. Fore more information refer to the literature listed in the beginning of Sect. 2.1.

2.1.3 Rigid Body Dynamics

The dynamic equations of a rigid robot with n DOF can be written as

$$\boldsymbol{M}(\boldsymbol{q})\ddot{\boldsymbol{q}} + \boldsymbol{C}(\boldsymbol{q}, \dot{\boldsymbol{q}})\dot{\boldsymbol{q}} + \boldsymbol{g}(\boldsymbol{q}) = \boldsymbol{\tau} + \boldsymbol{\tau}_{\text{ext}}. \tag{2.6}$$

The symmetric and positive definite inertia matrix $\boldsymbol{M}(\boldsymbol{q}) \in \mathbb{R}^{n\times n}$ depends on the joint configuration[3] $\boldsymbol{q} \in \mathbb{R}^n$. Gravity effects are contained in $\boldsymbol{g}(\boldsymbol{q}) = (\partial V_{\text{g}}(\boldsymbol{q})/\partial \boldsymbol{q})^T \in \mathbb{R}^n$, where $V_{\text{g}}(\boldsymbol{q})$ denotes the gravity potential. Coriolis/centrifugal forces and torques are represented by $\boldsymbol{C}(\boldsymbol{q}, \dot{\boldsymbol{q}})\dot{\boldsymbol{q}} \in \mathbb{R}^n$. The generalized forces[4] $\boldsymbol{\tau} \in \mathbb{R}^n$ describe the control inputs. Generalized external forces are denoted by $\boldsymbol{\tau}_{\text{ext}} \in \mathbb{R}^n$. The matrix $\boldsymbol{C}(\boldsymbol{q}, \dot{\boldsymbol{q}})$ is not unique in general, but it can be chosen according to the Christoffel symbols [MLS94] such that it complies with the relation

$$\dot{\boldsymbol{M}}(\boldsymbol{q}, \dot{\boldsymbol{q}}) = \boldsymbol{C}(\boldsymbol{q}, \dot{\boldsymbol{q}}) + \boldsymbol{C}(\boldsymbol{q}, \dot{\boldsymbol{q}})^T \tag{2.7}$$

which is in turn equivalent to the skew symmetry of $\dot{\boldsymbol{M}}(\boldsymbol{q}, \dot{\boldsymbol{q}}) - 2\boldsymbol{C}(\boldsymbol{q}, \dot{\boldsymbol{q}})$. This property is crucial for showing passivity of (2.6) w. r. t. input $(\boldsymbol{\tau} + \boldsymbol{\tau}_{\text{ext}})$ and output $\dot{\boldsymbol{q}}$ and the total energy $\frac{1}{2}\dot{\boldsymbol{q}}^T \boldsymbol{M}(\boldsymbol{q})\dot{\boldsymbol{q}} + V_{\text{g}}(\boldsymbol{q})$ as the storage function. This representation of the Coriolis/centrifugal matrix will be used by default in this book.

2.2 Compliant Motion Control of Robotic Systems

Featuring compliant behavior is an important requirement in many robotic applications. Consider task execution in unknown, dynamic, and unstructured environments such as households, or the cooperation of humans and robots in the same workspace. Whenever a physical contact between the robot and its environment occurs, the interaction behavior should be compliant or at least properly specified in terms of forces and torques. Two fundamental approaches exist to realize compliance: active control and the use of passive elements such as mechanical springs. Only the controlled compliance will be addressed in this book, whereas passive compliance is a matter of construction and mechanical design of the robot.

In the seminal work of Hogan [Hog85], the nature of physical systems is described from the environment point of view. They appear either as *admittances* accepting effort input (force) and yielding flow output (motion) or *impedances* accepting flow

[3]Positions for prismatic joints and angles for revolute joints.

[4]Forces for prismatic joints and torques for revolute joints.

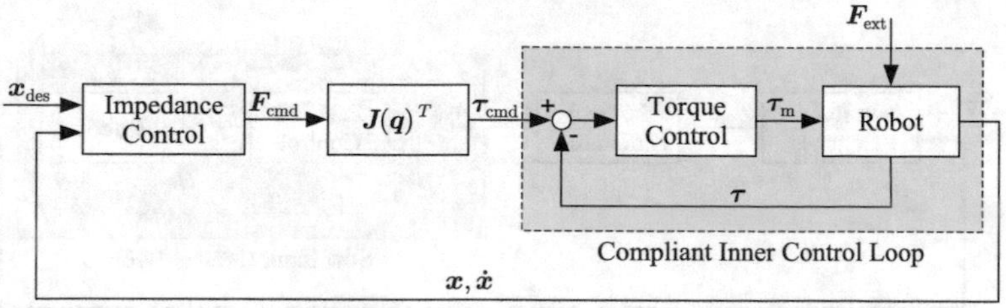

Fig. 2.2 Impedance control in the operational space $x \in \mathbb{R}^m$ with $m < n$

input (motion) and yielding effort output (force). His second fundamental statement is that dynamic interaction between physical systems cannot be controlled by exclusively commanding the position *or* the force. A controller has to incorporate the relation between these port variables as well. In this section the two classical approaches of impedance and admittance control for active compliance are briefly recapitulated. Especially impedance control [ASOH07, OASKH08] is of major importance in this monograph since the theory and the implementation of various impedance-based methods are addressed.

2.2.1 Impedance Control

The goal of impedance control is to alter the mechanical impedance of the robot, that is, the mapping from (generalized) velocities to (generalized) forces [Ott08]. Since the environment can be physically described as an admittance [Hog85] that maps forces to velocities, impressing an impedance behavior on the manipulator is a proper choice to define the interaction behavior in contact.

Figure 2.2 depicts the impedance control regulation case with setpoint $\boldsymbol{x}_{\text{des}}$ (des: desired) in the operational space $\boldsymbol{x} \in \mathbb{R}^m$ with $m < n$, for example in the Cartesian coordinates of the end-effector. This desired value is sometimes also called the virtual equilibrium. It is reached in the case of free motion, i.e. in the absence of external forces. By feeding the robot motion back, one computes the necessary force $\boldsymbol{F}_{\text{cmd}}$ (cmd: command) to implement the prescribed impedance. A kinematic mapping via the transpose of the Jacobian matrix $\boldsymbol{J}(\boldsymbol{q}) = \partial \boldsymbol{x} / \partial \boldsymbol{q}$ yields the required torque $\boldsymbol{\tau}_{\text{cmd}}$. The inner control loop realizes this torque by feedback of $\boldsymbol{\tau}$ under the influence of an external force $\boldsymbol{F}_{\text{ext}} \in \mathbb{R}^m$.[5] The impedance causality is $\dot{\boldsymbol{x}} \rightarrow \boldsymbol{F}_{\text{ext}}$, so repositioning the robot in the operational space results in forces acting on the environment.

When considering the overall structure of the impedance control in Fig. 2.2, one can conclude that the inner torque control loop is compliant while the outer loop

[5] In general, an external torque $\boldsymbol{\tau}_{\text{ext}} \in \mathbb{R}^n$ acts on the robot. If the contact with the environment is closed in the operational space, e.g. at the end-effector with coordinates $\boldsymbol{x}$, the force $\boldsymbol{F}_{\text{ext}} \in \mathbb{R}^m$ with $m < n$ is sufficient to describe the external load.

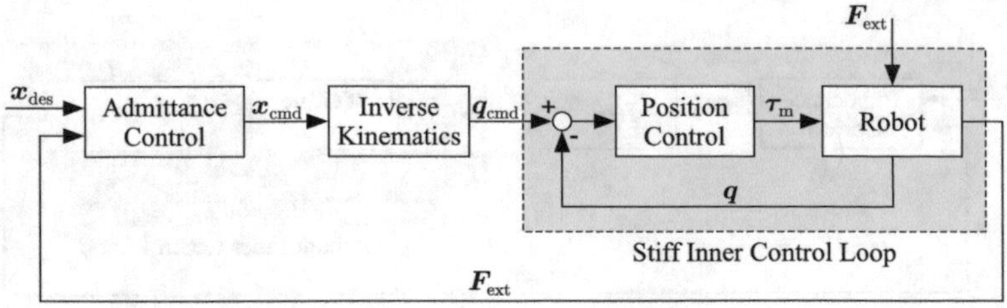

Fig. 2.3 Admittance control in the operational space $x \in \mathbb{R}^m$ with $m < n$

increases the stiffness of the complete system. The torque feedback in the inner loop allows for good contact behavior for small and medium stiffness in the impedance law. However, further increasing the stiffness will ultimately destabilize the system. Another characteristic of the impedance is the absence of an integrator. The controller has basically a PD structure. The inevitable steady-state error $\boldsymbol{x}_{des} - \boldsymbol{x}$ for $t \to \infty$ in the presence of model uncertainties or external loads can be reduced by increasing the stiffness. While the regulation case is depicted here, the tracking performance in case of a desired trajectory $\boldsymbol{x}_{des}(t)$ can be improved by adding a feed-forward term taking $\ddot{\boldsymbol{x}}_{des}(t)$ and the corresponding reflected inertia into account.

Compliance control is a special case of impedance regulation control, where the focus is put on the realization of a desired contact stiffness and damping [Ott08]. Compared to approaches based on the classical OSF [Kha87], where the perceived inertia is actively modified, the natural inertia of the robot is preserved in compliance control. The main advantage of the method is that the feedback of the generalized external forces is not required, which is beneficial in terms of robustness, availability of measurements, and the complexity of the implementation [OKN08].

2.2.2 Admittance Control

A mechanical admittance is the inverse of a mechanical impedance, that is, the mapping from (generalized) forces to (generalized) velocities. In compliant admittance control of robots, one employs a position or velocity controller in combination with explicit measurement and feedback of the generalized external forces. Figure 2.3 illustrates the implementation of such an admittance with joint position control interface. A typical example is Cartesian admittance control, where the external forces $\boldsymbol{F}_{ext}$ are measured at the tip of the end-effector by a six-axis force-torque sensor. As a result, compliance will only be achieved at the end-effector after this sensor, whereas the structure of the robot will react stiff in case of physical interaction.[6]

[6]One can also implement a compliant admittance controller on joint level via joint torque sensing, which would then lead to compliance along the entire manipulator.

The admittance subsystem yields a position $\boldsymbol{x}_{\text{cmd}}$ or velocity $\dot{\boldsymbol{x}}_{\text{cmd}}$ to be commanded. Then an inverse kinematics algorithm has to be employed to resolve the kinematic redundancy for $m < n$. The computed reference commands $\boldsymbol{q}_{\text{cmd}}$ or $\dot{\boldsymbol{q}}_{\text{cmd}}$, respectively, are realized in the inner position or velocity control loop. Since the admittance causality is $\boldsymbol{F}_{\text{ext}} \rightarrow \dot{\boldsymbol{x}}_{\text{cmd}}$, external forces result in a repositioning of the manipulator.

When considering the overall structure of the admittance control in Fig. 2.3, one can conclude that the inner position control loop is stiff, while the outer loop is responsible for the compliance. The inner-loop controller allows for high positioning accuracy, especially for medium and high stiffness. However, decreasing the stiffness will ultimately destabilize the system, which is a direct result of the admittance causality. Another restriction on its compliant contact behavior is the non-collocated feedback of the external forces in cases such as the Cartesian admittance control discussed above. The admittance-controlled system may involve substantial dynamics between actuator and sensor, which can lead to contact instability [CH89]. Admittance control is frequently used in robots which do not provide joint torque sensing or direct motor current interfaces [Ott08].

2.3 Humanoid Robot Rollin' Justin

The experiments in this book are mainly carried out on the wheeled humanoid robot Rollin' Justin[7] [OEF+06, BOW+07, BWS+09], see Fig. 2.4 (right). In Sect. 2.3.1, its hardware is presented as well as the underlying design principles. Section 2.3.2 summarizes several modeling assumptions that have to be made with regard to the implementation of the methods developed in this book.

2.3.1 Design and Hardware

The concept of Rollin' Justin is based on the principles of modularity and integrated design. In order to pass standard doorways, the overall width of the robot can be reduced to about 0.9 m by adjusting the upper body configuration and retracting the wheels. In terms of workspace, the robot is able to reach the floor as well as objects up to a height of about 2.7 m. The robot has an anthropomorphic structure which facilitates the operation in human environment where furniture, tools, and objects are optimized for the human anatomy. Another key feature of Rollin' Justin is that the robot can be operated without any cables that would restrict the mobility. It is equipped with a battery and all electronic components and computers are located onboard. Via WLAN and sensor feedback (speech recognition, visual information,

[7] The upper body of Justin was finished *just in* time for its first public presentation at the AUTOMATICA trade fair 2006 in Munich [OEF+06].

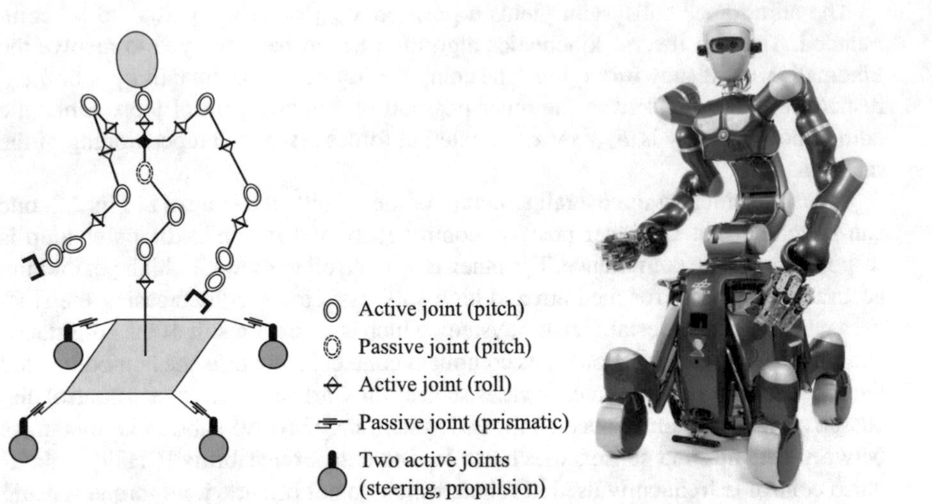

Fig. 2.4 Schematic (*left*) and picture (*right*) of the humanoid robot Rollin' Justin. The hands (12 actuated DOF each) are not specified in the sketch

force feedback), the robot has a permanent interface to the user and the environment during autonomous operation.

The arms are slightly modified versions of the DLR lightweight robots (LWR) of generation III [HSAS+02] with a total weight of 14 kg each. All electronic components are integrated in the arms. As with the human archetype, the arm has seven DOF. They are arranged in a roll-pitch-roll-pitch-roll-pitch-pitch order as it can be seen in the sketch in Fig. 2.4 (left). A payload of 15 kg can be lifted during slow motions and about 7 kg can be handled at maximum velocity. The hands of Rollin' Justin are the DLR Hands II [BGLH01] with four fingers and three actuated DOF per finger. An additional actuator has been integrated in the palm to reconfigure the alignment of the thumb, dependent on the application (power grasp, precision grasp). The head of Rollin' Justin is a pan-tilt unit that is equipped with several sensors, e.g. cameras for stereo vision and scene analysis or an inertial measurement unit for equilibrioception. The torso of Rollin' Justin has three actuated DOF and a kinematically coupled fourth one. The whole upper body weighs about 45 kg. Except for the two neck joints, all actuated upper body joints are equipped with link-side torque sensors as well as position sensors. This full state feedback makes it possible to implement various control techniques. The joint torque controller operates at a sampling rate of 3 kHz and the main control loop[8] runs at 1 kHz.

The mobile base with about 150 kg contains computers, battery, electronics and so forth [BWS+09]. Rollin' Justin has a variable footprint thanks to its extendable legs. A parallel mechanism ensures that the height of the platform remains unchanged.

[8]The main control loop contains all algorithms above the joint level such as Cartesian impedance control, self-collision avoidance, or online inverse kinematics.

Table 2.1 Actuated degrees of freedom and available control interfaces on Rollin' Justin

Subsystem	DOF	Control interface
Torso	3	Torque, position
Arms	2×7	Torque, position
Hands	2×12	Torque, position
Neck	2	Position
Platform and legs	8	Position, velocity
Total sum	51	

The leg extension DOF are not individually actuated, a reconfiguration is subject to steering and motion of the respective wheel. The leg lengths can also be locked mechanically. Due to the nonholonomy a dynamic feedback linearization is applied to move the platform [GFASH09]. This kinematic control method allows to realize arbitrary motions in the two translational directions forward/backward and left/right, and the rotation about the vertical axis. All actuated DOF of Rollin' Justin are summarized in Table 2.1.

2.3.2 *Modeling Assumptions*

Several assumptions concerning Rollin' Justin have to be made so that the approaches in this book can be applied to the robot.

Assumption 1 *The motors can be considered as ideal torque sources.*

The electrical time constants of the motors are sufficiently smaller than the mechanical ones. Therefore, one can neglect the electrical dynamics and assume ideal torque sources [Wim12, Ott08].

Assumption 2 *The reduced apparent motor inertia of the torque-controlled manipulator appears rigidly connected to the link inertia.*

The assumption is based on a singular perturbation argumentation applied to the flexible-joint model with large joint stiffness [Ott08, WO12, Wim12] and includes the so-called "inertia shaping" (downscaling of the apparent motor inertia via torque feedback) [OASK+04, ASOH04]. A fast time-scale inner torque controller is embedded in rigid body dynamics of slow time-scale. The apparent link inertia is modified by active control and the singular perturbation argumentation allows to neglect the dynamics between motor and link, resulting in a direct torque input available in the link dynamics as in (2.6).

Assumption 3 *The robot structure is rigid. Motions are restricted to the joints.*

The robot has a lightweight structure and is flexible in the links. Nevertheless, this link flexibility is negligible compared to the joint flexibility which originates from the Harmonic Drive gears and the strain-gauge-based torque sensors. Therefore, a flexible-joint model with concentrated elasticity in the joints can be assumed instead of an infinite-dimensional elastic-link robot model [Ott08].

Assumption 4 *The joint stiffness originating from the Harmonic Drive gears and the torque sensors in the joints is sufficiently high, such that the use of the motor positions instead of link positions for the kinematics does not lead to any noteworthy errors. The only exception concerns gravitational effects.*

Although the joint stiffness is high, gravity leads to deflections between motor and link. In order to compensate for gravity forces and torques properly, one has to take that flexibility into account. That can be realized by employing a *static equivalent of the link position* in the gravity model, which only depends on the motor position [OASK+04, ASOH07]. In order to remain in a passivity-based framework with collocated feedback, the motor position is used in the feedback controller. Once gravity is compensated as described, the dynamics between motor and link are neglected so that the motor positions (instead of the link positions) can be used for any link-side-dependent task definition or control task.

Assumption 5 *The rigid body dynamics* (2.6) *approximate the equations of motion of the flexible-joint model of Rollin' Justin, where the motor positions and the motor torques can be used instead of* $\boldsymbol{q}$ *and* $\boldsymbol{\tau}$*, respectively.*

The assumption of a direct torque input on link side is made possible by Assumptions 1 and 2, the assumption of rigid bodies is validated by Assumption 3, and the use of motor positions as a substitute for link positions is covered by Assumption 4.

Chapter 3
Control Tasks Based on Artificial Potential Fields

Robots with a large number of actuated DOF are able to perform several control tasks simultaneously. Holding a glass of water in a particular position and orientation in space, for example, requires six DOF. If there are more DOF available, this remaining kinematic redundancy can be used to pursue further important objectives such as collision avoidance, observation of the environment, or pose optimization for increased energy efficiency. In general, these control tasks have a *reactive* nature, i.e. the robot is able to locally react on disturbances, unmodeled dynamics, and unpredictable environments to achieve the goals in real time [DWASH12b]. Certainly the most frequently used method to define a reactive control task is to apply artificial attractive or repulsive potential fields [Kel29, Kha86] and to use their gradients as control inputs $\boldsymbol{\tau} \in \mathbb{R}^n$. In the general case, one can formulate

$$\boldsymbol{\tau} = -\nabla V(\boldsymbol{q}) = -\left(\frac{\partial V(\boldsymbol{q})}{\partial \boldsymbol{q}}\right)^T, \tag{3.1}$$

where $\nabla V(\boldsymbol{q})$ is the gradient of the artificial potential $V(\boldsymbol{q}) \in \mathbb{R}_0^+$ and $\boldsymbol{\tau}$ follows the gradient descent. Artificial potentials are intuitive and simple to parameterize because the controller gains have a direct relation to the physical world, e.g. the potential stiffness which can be interpreted as the stiffness of a virtual spring as in Fig. 1.2.

In the last decades, numerous methods have been developed based on these principles, ranging from control tasks to usage within planning algorithms [SK05, LCCF11, DWASH12b, BK02, BHG10]. In the context of dexterous manipulation, especially Cartesian end-effector control [Kha87, ASOH07], manipulator singularity avoidance [Ott08], and avoidance of mechanical end stops [MCR96] are of high importance, since they are essential components in almost every whole-body control framework.

In this chapter, the set of standard control tasks, as outlined above, is extended by particular applications which are not satisfactorily covered by the state of the art:

A. Dietrich, *Whole-Body Impedance Control of Wheeled Humanoid Robots*,
Springer Tracts in Advanced Robotics 116, DOI 10.1007/978-3-319-40557-5_3

In Sect. 3.1, a reactive self-collision avoidance algorithm is proposed, that generates repulsive forces between potentially colliding links of the robot. Especially when a robot has many DOF, the problem of self-collisions becomes a crucial issue in whole-body control. The singularity-free control of wheeled mobile platforms is addressed in Sect. 3.2. When the instantaneous center of rotation approaches the steering axis of a wheel, then the required steering velocity goes to infinity. This critical situation is avoided by repelling the instantaneous center of rotation from the steering axes by means of artificial potential fields. In Sect. 3.3, the kinematic and dynamic workspace of tendon-coupled torsos is treated using the example of Rollin' Justin. Repulsive forces are generated to fully exploit the workspace of its torso. Section 3.4 closes the chapter with a brief recapitulation of classical objectives in reactive, potential-field-based control, which will also be applied in the context of whole-body control later. There, the controllers of this chapter will be given priorities and applied to the robot simultaneously. Redundancy resolutions (Chap. 4) will be exploited to realize this hierarchical stack of whole-body control tasks.

3.1 Self-Collision Avoidance

The large number of DOF of humanoid robots increases the complexity in terms of self-collisions. Using only planning algorithms to prevent collisions is not sufficient if compliant physical interaction is considered, where the configuration of the robot may change significantly. Then the robot must be able to detect critical situations and react in real time [KNK+02, HASH08]. The classical approach is to apply repulsive potential fields [Kha86], both for the treatment of self-collisions and collisions with external objects. In [SGJG10], Sugiura et al. generate repulsive forces between potentially colliding body links of ASIMO and transform them into the corresponding joint motions via an admittance in order to access the velocity control interface. An alternative approach has been implemented on the HRP-2 humanoid robot by Stasse et al. [SEM+08]. Based on cost functions, collisions are avoided by kinematic control.

Compared to the majority of the state-of-the-art approaches, the controller presented here (Fig. 3.1) does not work on a kinematic level but it commands joint torques. Thus, it is better suited for physical interaction and typical environments

(a) **(b)** **(c)**

Fig. 3.1 Public presentation of the self-collision avoidance at the AUTOMATICA trade fair 2010 in Munich (Germany). **a** Before potential self-collision. **b** Active repulsion to avoid self-collision. **c** After successful self-collision avoidance

of service robots. The approach is generic and can be applied to any robotic system. Besides the generation of repulsive forces between close links, it also includes a configuration-dependent damping design for systematic dissipation of kinetic energy in a well-directed manner. The algorithm deals with a large number of potentially colliding body segments simultaneously while being real-time capable. Preliminary work has been done by De Santis et al. [DSASO+07]. They have already utilized the torque control interface, but specific damping has not been considered yet. Moreover, the available collision models were insufficient to describe the complex robot geometry. The following sections are based on [DWT+11, DWASH12a].

3.1.1 Geometric Collision Model

The self-collision avoidance (sca) generates forces between potentially colliding links. In the first step, the respective links and points on the surface have to be determined. Since multiple collisions are possible at the same time, one has to consider a sufficiently large number of collision pairs $n_{\text{sca}} \in \mathbb{N}$. Each pair consists of two contact points $\boldsymbol{x}_i, \boldsymbol{x}_j \in \mathbb{R}^3$ on potentially colliding links. Such a *contact point pair* is unambiguously identified by its indices (i, j).

The algorithm combines a compact, numerically efficient volume representation and a standard distance computation algorithm for convex hulls [GJK88]. In each control cycle, the geometric collision model is updated: All volumes are transformed into the world frame and the distances as well as the contact points are determined. Collision pairs which cannot collide are excluded to reduce the computation time of the algorithm. An example of such a geometric model is given in Fig. 3.2. The bounding volumes are designed to be as tight as possible while being represented by simple geometric bodies such as spheres or rounded cylinders. The volumes are spherically extended convex hulls, so-called *swept sphere volumes*. The model in Fig. 3.2 uses 78 points and 28 radii. For a more detailed description, the reader may refer to [TBF11]. The design of the bounding volumes is a compromise between accuracy (tight hulls) and numerical efficiency (simple geometries).

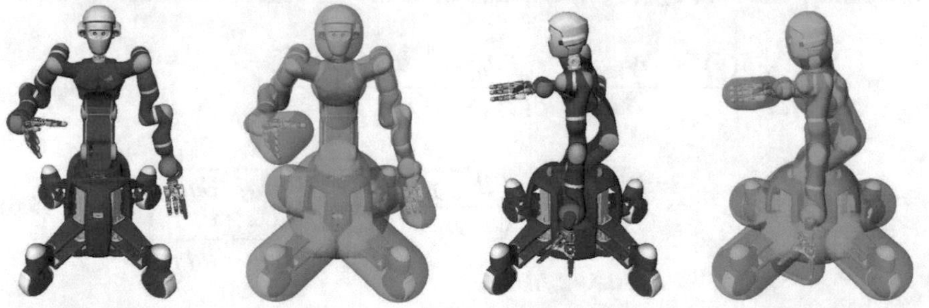

Fig. 3.2 Geometric collision model of Rollin' Justin consisting of 28 bounding volumes (left arm: 8, right arm: 8, mobile base: 5, torso: 4, head: 2, floor: 1). The volumes are spheres, rounded cylinders, and bodies obtained by unrolling spheres on triangles

3.1.2 Repulsive Potential

The control action $\boldsymbol{\tau}_{\text{sca}} \in \mathbb{R}^n$ for self-collision avoidance can be expressed by

$$\boldsymbol{\tau}_{\text{sca}} = -\left(\frac{\partial V_{\text{sca}}(\boldsymbol{q})}{\partial \boldsymbol{q}}\right)^T - \boldsymbol{D}(\boldsymbol{q})\dot{\boldsymbol{q}}, \tag{3.2}$$

where the positive definite damping matrix $\boldsymbol{D}(\boldsymbol{q}) \in \mathbb{R}^{n\times n}$ provides the means to specifically dissipate kinetic energy. The derivation of this configuration-dependent matrix will be presented in Sect. 3.1.3. The term $V_{\text{sca}}(\boldsymbol{q}) \in \mathbb{R}_0^+$ denotes the total potential energy of all n_{sca} potential fields applied to the geometric model according to

$$V_{\text{sca}}(\boldsymbol{q}) = \sum_{i=1}^{n_{\text{sca}}} V_{\text{sca},(i,h(i))}(\boldsymbol{q}). \tag{3.3}$$

Therein, each potential $V_{\text{sca},(i,h(i))}(\boldsymbol{q}) \in \mathbb{R}_0^+$ refers to the contact point pair defined by point i and corresponding point j with the assignment $h : i \mapsto j$ for $i = 1 \ldots n_{\text{sca}}$. The distance $d_{(i,j)} \in \mathbb{R}_0^+$ between two contact points $\boldsymbol{x}_i, \boldsymbol{x}_j$ is defined as

$$d_{(i,j)} = ||\boldsymbol{x}_i^{\mathcal{C}(i)} - \boldsymbol{x}_j^{\mathcal{C}(i)}|| = ||\boldsymbol{x}_j^{\mathcal{C}(j)} - \boldsymbol{x}_i^{\mathcal{C}(j)}||. \tag{3.4}$$

The superscript describes the coordinate frame of the link on which the indicated contact point lies. Such a point pair $(\boldsymbol{x}_i^{\mathcal{C}(i)}, \boldsymbol{x}_j^{\mathcal{C}(i)})$ is illustrated in the three-dimensional example of Fig. 3.3. The repulsive forces $F_{\text{sca}}(d_{(i,j)}) \in \mathbb{R}_0^+$ are perpendicular to the surfaces of the links and point in the directions $\pm\boldsymbol{e}_i$ with

$$\boldsymbol{e}_i = \frac{\boldsymbol{x}_j^{\mathcal{C}(i)} - \boldsymbol{x}_i^{\mathcal{C}(i)}}{d_{(i,j)}}. \tag{3.5}$$

As of now, the superscript $\mathcal{C}(i)$ will be omitted in the notations for the sake of readability. The term $F_{\text{sca}}(d_{(i,j)})$ is contained in the gradient of the potential (3.2):

$$\begin{aligned} \frac{\partial V_{\text{sca},(i,j)}(\boldsymbol{q})}{\partial \boldsymbol{q}} &= \frac{\partial V_{\text{sca},(i,j)}}{\partial d_{(i,j)}} \frac{\partial d_{(i,j)}}{\partial (\boldsymbol{x}_i^T, \boldsymbol{x}_j^T)^T} \frac{\partial (\boldsymbol{x}_i^T, \boldsymbol{x}_j^T)^T}{\partial \boldsymbol{q}} \\ &= \underbrace{\frac{\partial V_{\text{sca},(i,j)}}{\partial d_{(i,j)}}}_{-F_{\text{sca}}(d_{(i,j)})} \begin{pmatrix} \frac{\partial d_{(i,j)}}{\partial \boldsymbol{x}_i} & \frac{\partial d_{(i,j)}}{\partial \boldsymbol{x}_j} \end{pmatrix} \begin{pmatrix} \frac{\partial \boldsymbol{x}_i}{\partial \boldsymbol{q}_i} & \frac{\partial \boldsymbol{x}_i}{\partial \boldsymbol{q}_j} \\ \frac{\partial \boldsymbol{x}_j}{\partial \boldsymbol{q}_i} & \frac{\partial \boldsymbol{x}_j}{\partial \boldsymbol{q}_j} \end{pmatrix}. \end{aligned} \tag{3.6}$$

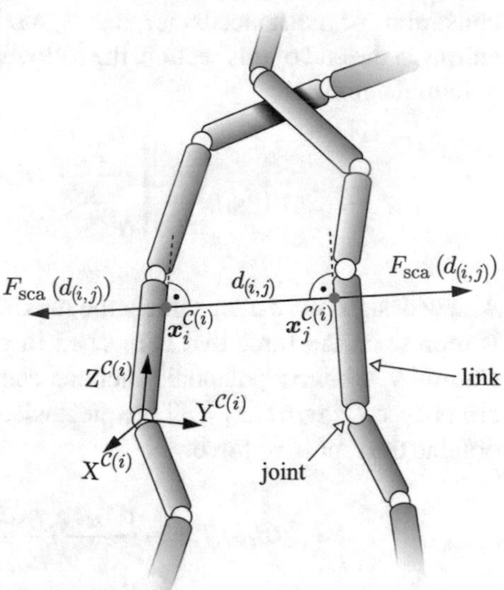

Fig. 3.3 Relations between two arbitrary contact points $\boldsymbol{x}_i^{C(i)}$ and $\boldsymbol{x}_j^{C(i)}$ and repulsive forces $F_{\text{sca}}(d_{(i,j)})$. The figure illustrates a 3-dimensional example

Herein, the vectors $\boldsymbol{q}_i$ and $\boldsymbol{q}_j$ denote the joint values which directly[1] affect the location of $\boldsymbol{x}_i$ and $\boldsymbol{x}_j$, respectively. In the example in Fig. 3.3, $\boldsymbol{q}_i$ and $\boldsymbol{q}_j$ describe the joint values of the left and the right manipulator, respectively. Notice that in general the contact points may have the same base of the kinematic chain so that $\boldsymbol{q}_i$ and $\boldsymbol{q}_j$ have an intersection. Nonetheless, just the joints after the branch-off point are relevant. It follows from the multiplication of $\partial d_{(i,j)}/\partial(\boldsymbol{x}_i^T, \boldsymbol{x}_j^T)^T$ and $\partial(\boldsymbol{x}_i^T, \boldsymbol{x}_j^T)^T/\partial \boldsymbol{q}$ that just the principal block diagonal of $\partial(\boldsymbol{x}_i^T, \boldsymbol{x}_j^T)^T/\partial \boldsymbol{q}$ has influence on the result. The other multiplications are zero because the factors are always orthogonal. As an example, let us consider $\partial d_{(i,j)}/\partial \boldsymbol{x}_j$ and $\partial \boldsymbol{x}_j/\partial \boldsymbol{q}_i$ in the context of Fig. 3.3. The direction of $\partial d_{(i,j)}/\partial \boldsymbol{x}_j$ is orthogonal to the surface of the link on which $\boldsymbol{x}_j$ is lying, whereas $\boldsymbol{q}_i$ is only able to let $\boldsymbol{x}_j$ move on this surface (indirect influence). Hence, (3.6) can be simplified to

$$\frac{\partial V_{\text{sca},(i,j)}(\boldsymbol{q})}{\partial \boldsymbol{q}} = -F_{\text{sca}}(d_{(i,j)}) \underbrace{\left(\frac{\partial d_{(i,j)}}{\partial \boldsymbol{x}_i} \frac{\partial \boldsymbol{x}_i}{\partial \boldsymbol{q}_i} \;\; \frac{\partial d_{(i,j)}}{\partial \boldsymbol{x}_j} \frac{\partial \boldsymbol{x}_j}{\partial \boldsymbol{q}_j} \right)}_{\left(\boldsymbol{J}_i(\boldsymbol{q}),\ \boldsymbol{J}_j(\boldsymbol{q})\right)} . \tag{3.7}$$

The right part in (3.7) describes the mapping from joint space to collision space, i.e. the Jacobian matrices $\boldsymbol{J}_i(\boldsymbol{q})$ and $\boldsymbol{J}_j(\boldsymbol{q})$, respectively. These mappings will be used in the damping design in Sect. 3.1.3.

The repulsive potentials $V_{\text{sca},(i,j)}(\boldsymbol{q})$ are zero at a specified distance $d_{(i,j)} = d_0$. That is due to the requirement of the self-collision avoidance being a unilateral

[1] Indirect influence implies the motion of the point on the surface of the link due to the motion of the corresponding contact point partner.

constraint. At a distance larger than d_0, no torque will be applied. In the implementations at the end of this section, the following choice has been made for the artificial potential definition:

$$V_{\mathrm{sca},(i,j)}(d_{(i,j)}) = \begin{cases} -\frac{F_{\max}}{3d_0^2}\left(d_{(i,j)} - d_0\right)^3 & \text{for } d_{(i,j)} \leq d_0 \\ 0 & \text{for } d_{(i,j)} > d_0 \end{cases}. \tag{3.8}$$

A new design parameter, namely the maximum force $F_{\max} \in \mathbb{R}^+$, must be specified. It represents the force that is reached in case of collision contact, i.e. $d_{(i,j)} = 0$. Naturally, arbitrary potential functions can be used instead of (3.8) as long as they are of type C^2 w.r.t. $d_{(i,j)}$. If the piecewise defined function (3.8) is employed, one obtains the repulsive force

$$F_{\mathrm{sca}}(d_{(i,j)}) = -\frac{\partial V_{\mathrm{sca},(i,j)}(d_{(i,j)})}{\partial d_{(i,j)}} \tag{3.9}$$

$$= \begin{cases} \frac{F_{\max}}{d_0^2}\left(d_{(i,j)} - d_0\right)^2 & \text{for } d_{(i,j)} \leq d_0 \\ 0 & \text{for } d_{(i,j)} > d_0 \end{cases}. \tag{3.10}$$

The reason behind the requirement of C^2 for $V_{\mathrm{sca},(i,j)}(d_{(i,j)})$ is that the damping force in Sect. 3.1.3 will directly depend on the local potential stiffness $\partial^2 V_{\mathrm{sca},(i,j)}(d_{(i,j)})/\partial d_{(i,j)}^2$, and that force is required to be continuous in order to guarantee a continuous control law on torque level.

The two parameters d_0 and $F_{\max}$ have to be set in order to uniquely define the potentials in (3.8). The following list shows a selection of possible design criteria:

- The maximum local stiffness $\left(\partial F_{\mathrm{sca}}(d_{(i,j)})/\partial d_{(i,j)}\right)\big|_{d_{(i,j)}=0}$ can be limited to a feasible value dependent on the hardware, sample time, and bandwidth of the torque control.
- A rough estimation for the worst case $F_{\max}$ or joint torque can be made, i.e. one contact point pair is supposed to avoid a self-collision in a critical situation.
- A rough estimation of the maximum kinetic energy in the robot links can be utilized to design the total energy storage of the artificial potentials.

However, since multiple contact point pairs and configuration-dependent relations between repulsive forces and joint torques are considered, such a design is not trivial in general.

3.1.3 Damping Design

The repulsion as described above will avoid collisions, but it will not dissipate the energy which is stored in the virtual elastic springs. As a result, the links will oscillate

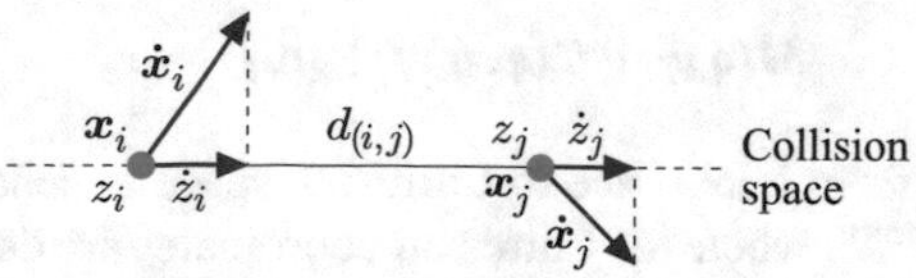

Fig. 3.4 Illustration of the projected motions of $\boldsymbol{x}_i$ and $\boldsymbol{x}_j$ in the direction of the collision. Positive directions are defined from $\boldsymbol{x}_i$ to $\boldsymbol{x}_j$. The new coordinates in the collision space are denoted by z_i and z_j

back and forth, alternately converting elastic energy to kinetic energy and converting it back to elastic energy. Active damping for energy dissipation has to be introduced to prevent such a situation.

The following damping design facilitates a systematic energy dissipation by specification of damping ratios $\xi \in \mathbb{R}^+$ for the contact point pairs. The first step consists of a coordinate transformation of the dynamic equations to the relevant operational space [Kha87], which is defined by the collision directions. That procedure is required for each contact point pair. The new coordinates are given by the projections z_i and z_j of $\boldsymbol{x}_i$ and $\boldsymbol{x}_j$ in the collision space (direction of the possible collision), see Fig. 3.4. Positive directions for both values are defined from $\boldsymbol{x}_i$ to $\boldsymbol{x}_j$.

Following that, a desired, standard rigid body robot differential equation [MLS94] can be set up:

$$\boldsymbol{M}_{(i,j)}(\boldsymbol{q})\begin{pmatrix}\ddot{z}_i\\ \ddot{z}_j\end{pmatrix} + \boldsymbol{C}_{(i,j)}(\boldsymbol{q},\dot{\boldsymbol{q}})\dot{\boldsymbol{q}} + \boldsymbol{g}_{(i,j)}(\boldsymbol{q}) = \boldsymbol{F}_{(i,j)}, \tag{3.11}$$

$$\boldsymbol{F}_{(i,j)} = -\boldsymbol{D}_{(i,j)}(\boldsymbol{q})\begin{pmatrix}\dot{z}_i\\ \dot{z}_j\end{pmatrix} - F_{\text{sca}}(d_{(i,j)})\begin{pmatrix}1\\ -1\end{pmatrix}. \tag{3.12}$$

Coriolis and centrifugal effects are represented by $\boldsymbol{C}_{(i,j)}(\boldsymbol{q},\dot{\boldsymbol{q}}) \in \mathbb{R}^{2\times 2}$, and gravity torques are expressed by $\boldsymbol{g}_{(i,j)}(\boldsymbol{q}) \in \mathbb{R}^2$. The inertia matrix $\boldsymbol{M}_{(i,j)}(\boldsymbol{q}) \in \mathbb{R}^{2\times 2}$ contains the reflected inertias at the contact points (i, j) in the direction of the collision and has the form

$$\boldsymbol{M}_{(i,j)}(\boldsymbol{q}) = \begin{pmatrix} m_i(\boldsymbol{q}) & 0\\ 0 & m_j(\boldsymbol{q})\end{pmatrix}. \tag{3.13}$$

The configuration-dependent damping matrix $\boldsymbol{D}_{(i,j)}(\boldsymbol{q}) \in \mathbb{R}^{2\times 2}$ is derived in the following. Since the damping ratios depend on the dynamics (3.11), the inertia (3.13) has to be determined first.

3.1.3.1 Reflected Inertia $m_i(\boldsymbol{q})$

In the following derivation, the known joint inertia matrix $\boldsymbol{M}(\boldsymbol{q})$ is to be transformed into $m_i(\boldsymbol{q})$. The scalar $m_j(\boldsymbol{q})$ can be obtained analogously. The general relation between joint torque $\boldsymbol{\tau}$ and joint acceleration $\ddot{\boldsymbol{q}}$ is

$$\boldsymbol{M}(\boldsymbol{q})\ddot{\boldsymbol{q}} + \boldsymbol{C}(\boldsymbol{q}, \dot{\boldsymbol{q}})\dot{\boldsymbol{q}} + \boldsymbol{g}(\boldsymbol{q}) = \boldsymbol{\tau}. \tag{3.14}$$

The transformation from joint space to Cartesian space is denoted by the Jacobian matrix $\boldsymbol{J}_{\boldsymbol{x}_i,\boldsymbol{q}}(\boldsymbol{q}) \in \mathbb{R}^{3\times n}$, where the Cartesian coordinates are defined by an arbitrary point $\boldsymbol{x}_i^{\mathcal{C}(i)}(\boldsymbol{q}) \in \mathbb{R}^3$:

$$\dot{\boldsymbol{x}}_i^{\mathcal{C}(i)} = \boldsymbol{J}_{\boldsymbol{x}_i,\boldsymbol{q}}(\boldsymbol{q})\dot{\boldsymbol{q}}, \tag{3.15}$$

$$\boldsymbol{\tau} = \boldsymbol{J}_{\boldsymbol{x}_i,\boldsymbol{q}}(\boldsymbol{q})^T \boldsymbol{F}_{\boldsymbol{x}_i}. \tag{3.16}$$

The term $\boldsymbol{F}_{\boldsymbol{x}_i} \in \mathbb{R}^3$ describes an external force applied at $\boldsymbol{x}_i^{\mathcal{C}(i)}$. In the next step, the projection in the direction of the collision $\boldsymbol{e}_i$ has to be considered. The mapping

$$\boldsymbol{J}_{z_i,\boldsymbol{x}_i}(\boldsymbol{e}_i) = \boldsymbol{e}_i^T \tag{3.17}$$

relates motions in the Cartesian directions of the respective contact point to motions along $\boldsymbol{e}_i$. The combination of (3.15) and (3.17) delivers

$$\dot{z}_i = \boldsymbol{J}_{z_i,\boldsymbol{x}_i}(\boldsymbol{e}_i)\dot{\boldsymbol{x}}_i^{\mathcal{C}(i)} = \underbrace{\boldsymbol{J}_{z_i,\boldsymbol{x}_i}(\boldsymbol{e}_i)\boldsymbol{J}_{\boldsymbol{x}_i,\boldsymbol{q}}(\boldsymbol{q})}_{\boldsymbol{J}_i(\boldsymbol{q})}\dot{\boldsymbol{q}} \tag{3.18}$$

with the row vector Jacobian $\boldsymbol{J}_i(\boldsymbol{q}) \in \mathbb{R}^{1\times n}$ relating the joint space to the collision space as introduced in (3.7). The velocity $\dot{z}_j$ of contact point $\boldsymbol{x}_j$ in the collision space can be derived analogously with $\boldsymbol{e}_j$ being oriented in the same direction as $\boldsymbol{e}_i$ according to Fig. 3.4. The total time derivative of (3.18) delivers the acceleration constraint

$$\ddot{z}_i = \boldsymbol{J}_i(\boldsymbol{q})\ddot{\boldsymbol{q}} + \dot{\boldsymbol{J}}_i(\boldsymbol{q}, \dot{\boldsymbol{q}})\dot{\boldsymbol{q}}, \tag{3.19}$$

and, taking account of (3.14) and (3.16),

$$\ddot{z}_i = \dot{\boldsymbol{J}}_i(\boldsymbol{q}, \dot{\boldsymbol{q}})\dot{\boldsymbol{q}} - \boldsymbol{J}_i(\boldsymbol{q})\boldsymbol{M}(\boldsymbol{q})^{-1}(\boldsymbol{C}(\boldsymbol{q}, \dot{\boldsymbol{q}})\dot{\boldsymbol{q}} + \boldsymbol{g}(\boldsymbol{q})) + \underbrace{\boldsymbol{J}_i(\boldsymbol{q})\boldsymbol{M}(\boldsymbol{q})^{-1}\boldsymbol{J}_i^T(\boldsymbol{q})}_{m_i(\boldsymbol{q})^{-1}} F_{z_i}. \tag{3.20}$$

The scalar force $F_{z_i} \in \mathbb{R}$ acts at z_i and accelerates the mass $m_i(\boldsymbol{q})$ in the collision space. From a computational point of view, the evaluation of $m_i(\boldsymbol{q})$ and $m_j(\boldsymbol{q})$ is not expensive because the last inversion refers to scalar. Inverting the joint inertia matrix $\boldsymbol{M}(\boldsymbol{q})$ has to be done only once per sample time, independent of the number of contact point pairs.

3.1.3.2 Damping Matrix $D_{(i,j)}(q)$

Since damping ratios are defined for linear dynamics and (3.11) is nonlinear, that issue has to be treated first. The linearization around the working point (denoted by superscript *) with $d^*_{(i,j)} = f(q^*)$ under the assumption of a quasi-static analysis, i.e. $\dot{z}^*_i = \dot{z}^*_j = 0$, delivers

$$M_{(i,j)}(q^*)\begin{pmatrix}\delta\ddot{z}_i\\ \delta\ddot{z}_j\end{pmatrix} + D_{(i,j)}(q^*)\begin{pmatrix}\delta\dot{z}_i\\ \delta\dot{z}_j\end{pmatrix} + K_{(i,j)}(q^*)\begin{pmatrix}\delta z_i\\ \delta z_j\end{pmatrix} = \mathbf{0}, \tag{3.21}$$

The local stiffness matrix $K_{(i,j)}(q^*) \in \mathbb{R}^{2\times 2}$ is

$$\begin{aligned} K_{(i,j)}(q^*) &= \begin{pmatrix}1\\-1\end{pmatrix} \left.\frac{\partial F_{\text{sca}}(d_{(i,j)})}{\partial d_{(i,j)}}\right|_{d_{(i,j)}=d^*_{(i,j)}} \cdot \frac{\partial d_{(i,j)}}{\partial (z_i\ z_j)} \\ &= \left.\frac{\partial^2 V_{\text{sca},(i,j)}(d_{(i,j)})}{\partial d^2_{(i,j)}}\right|_{d_{(i,j)}=d^*_{(i,j)}} \cdot \begin{pmatrix}1 & -1\\ -1 & 1\end{pmatrix}. \end{aligned} \tag{3.22}$$

Gravity effects are omitted in the linearized version of the dynamics (3.21) because a separate gravity compensation is applied in the overall control law. The local damping behavior is specified by $D_{(i,j)}(q^*) \in \mathbb{R}^{2\times 2}$. Knowledge of the inertia and stiffness parameters in (3.21) facilitates various methods from linear algebra theory in order to implement the desired damping ratios.[2] Here, the *Double Diagonalization* approach by Albu-Schäffer et al. [ASOFH03] is applied. The damping matrix can be formally written as

$$D_{(i,j)}(q^*) = \mathcal{D}\left(M_{(i,j)}(q^*), K_{(i,j)}(q^*), \xi\right), \tag{3.23}$$

where the damping ratio ξ is implemented for both directions. The algorithm will be computed and applied in each control cycle.

3.1.4 Control Design

If all n_{sca} contact point pairs are considered and equally weighted, the self-collision avoidance torques are

$$\tau_{\text{sca}} = \sum_{i=1}^{n_{\text{sca}}} \begin{pmatrix}J_i(q)\\ J_j(q)\end{pmatrix}^T \left(\begin{pmatrix}-F_{\text{sca}}(d_{(i,j)})\\ F_{\text{sca}}(d_{(i,j)})\end{pmatrix} - D_{(i,j)}(q^*)\begin{pmatrix}J_i(q)\\ J_j(q)\end{pmatrix}\dot{q}\right). \tag{3.24}$$

---

[2] Although the damping design is done for the linear system, it is still valid for the analysis of the nonlinear system as the dissipation term only appears in the time derivative of the Lyapunov function.

Table 3.1 Parameterization for the experiment on self-collision avoidance

F_{max}	d_0	n_{sca}	$\xi = \xi_1 = \xi_2$
25 N	0.15 m	35	0, 0.7, 1.0, 1.3

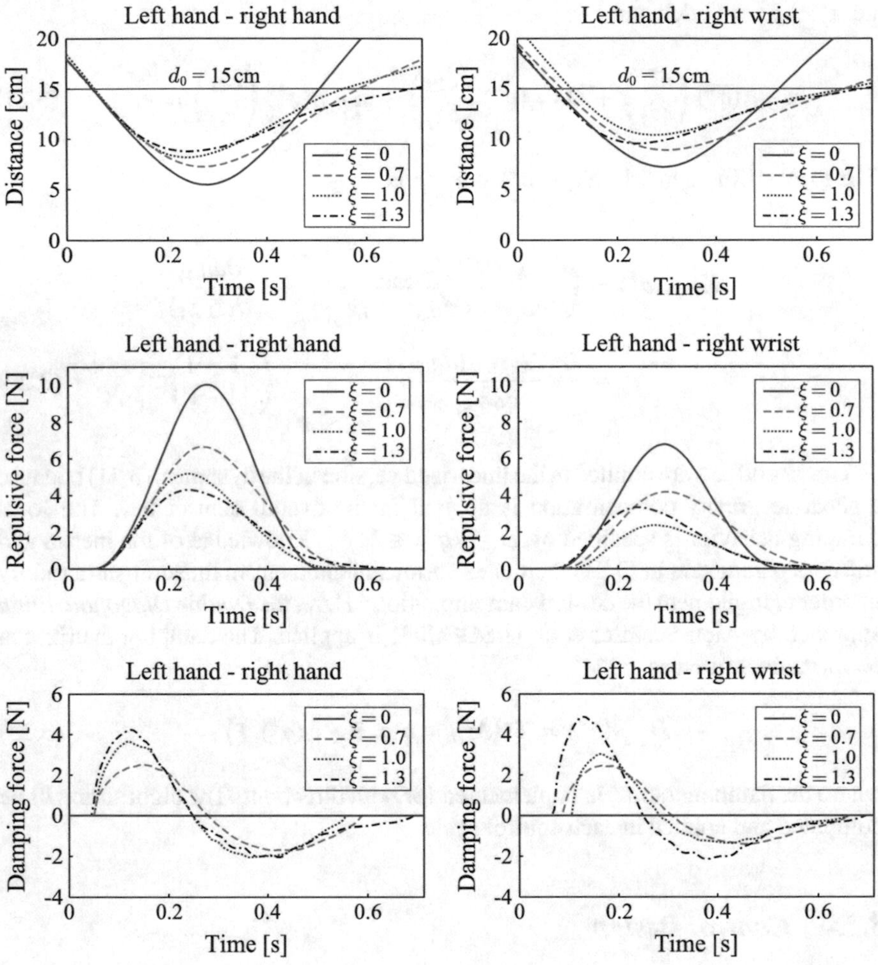

Fig. 3.5 Experiment on self-collision avoidance: repulsion between left hand and right hand/wrist for the undamped ($\xi = 0$), underdamped ($\xi = 0.7$), critically damped ($\xi = 1.0$), and overdamped ($\xi = 1.3$) system. The two contact point pairs are a selection of a total number of 14 active pairs

This superposition can generate local minima. In Chap. 4, it will be shown how a hierarchy among the individual repulsive potentials can be realized.

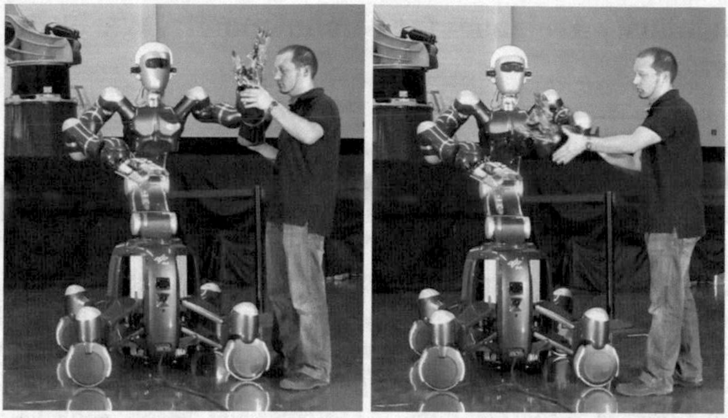

Fig. 3.6 Initial configuration (*left*) and snapshot during the experiment (*right*) on self-collision avoidance. Except for the left arm (seven DOF), all other joints are locked in order to facilitate repeatability of the experiment

3.1.5 Experiments

The performance of the self-collision avoidance is evaluated on Rollin' Justin using the parameters provided in Table 3.1. In the experiments, the geometric collision model from Fig. 3.2 is used. The distance computation algorithm is based on the formulation in [GJK88] and has been adapted to the system, see [TBF11] for more details. In order to facilitate real-time applicability, the collision model is calculated once per control cycle (1 ms) incorporating 302 pairs of links. The computing time lies between 0.3 and 0.4 ms on a standard Intel Core2Duo Processor T7400 (2.16 GHz).

In each run, a different damping ratio is applied. Throughout, the robot is controlled in gravity compensation mode. Except for the left arm (seven DOF), all other joints are locked. Figure 3.6 (left) shows the initial configuration. The user feeds kinetic energy into the system, throwing the left forearm onto the right arm (right). A total number of 14 potential fields are activated during the measurements including the arms, the torso, the mobile base, and the head. The most critical ones are shown in Fig. 3.5. They refer to the sets "left hand–right hand" (left column diagrams) and "left hand–right wrist" (right column diagrams). In all scenarios the user injects about the same amount of kinetic energy.[3] As expected, the penetration depths of the potential fields are significantly smaller while damping is active. Naturally, the returning velocities of the links are affected by the choice of ξ as it can be seen in the upper plots in Fig. 3.5.

[3]The energies which are absorbed by the most relevant potential field, i.e., "left hand–right hand", have a maximum difference of $<12\,\%$.

3.2 Singularity Avoidance for Nonholonomic, Wheeled Platforms

Controlling nonholonomic, mobile robots requires precise orientation of the wheels at each instant. Otherwise, high internal forces are generated, which stress the mechanical structure. One of the most common methods is based on the explicit use/command of the instantaneous center of rotation (ICR) [LNL+06, CPHV08, LQXX09]. However, singularities in the steering velocities of the wheels occur when the ICR crosses or comes close to one of the steering axes. One way to circumvent that problem is to use special wheels [WA99] or to apply constraints to the accessible velocity space in order to avoid singular regions [TDNM96]. Another method to avoid these problematic configurations is to consider them as obstacles and design repulsive potentials. This approach has been applied by Connette et al. in [CPHV09] and has proven successful. The method proposed here [DWASH11] is an extension of the latter kind in case of mobile platforms with variable footprint. Moreover, the concept does not permanently use the explicit representation of the ICR in contrast to [CPHV09]. That way, additional singularities based on the mathematical description of the ICR itself are avoided.

3.2.1 Instantaneous Center of Rotation

The instantaneous center of rotation (ICR) is the unique point $\boldsymbol{p}_{\text{icr}} \in \mathbb{R}^2$ (expressed here in platform body frame) around which the vehicle rotates at each instant. The ICR in Cartesian body coordinates can be expressed as

$$\boldsymbol{p}_{\text{icr}} = \begin{pmatrix} x_{\text{icr}} \\ y_{\text{icr}} \end{pmatrix} = \underbrace{\frac{v}{|\dot{\theta}|}}_{r_{\text{icr}}} \begin{pmatrix} \cos(\gamma) \\ \sin(\gamma) \end{pmatrix}, \tag{3.25}$$

$$v = \sqrt{\dot{x}^2 + \dot{y}^2}. \tag{3.26}$$

The graphical interpretation is given in Fig. 3.7. The numerator in (3.25) characterizes the absolute translational velocity and the denominator the absolute angular velocity of the platform center. The quotient defines the radius of curvature r_{icr} (distance from origin to ICR), whereas γ indicates the corresponding direction. The latter derives straightforwardly from the direction of motion (which is determined by $\dot{x}$ and $\dot{y}$) and the direction of rotation.[4]

$$\gamma = \arctan_2\left(\text{sgn}(\dot{\theta})\dot{x}, -\text{sgn}(\dot{\theta})\dot{y}\right). \tag{3.27}$$

[4]Concerning the order of arguments, the form $\arctan_2(y, x)$ is used here.

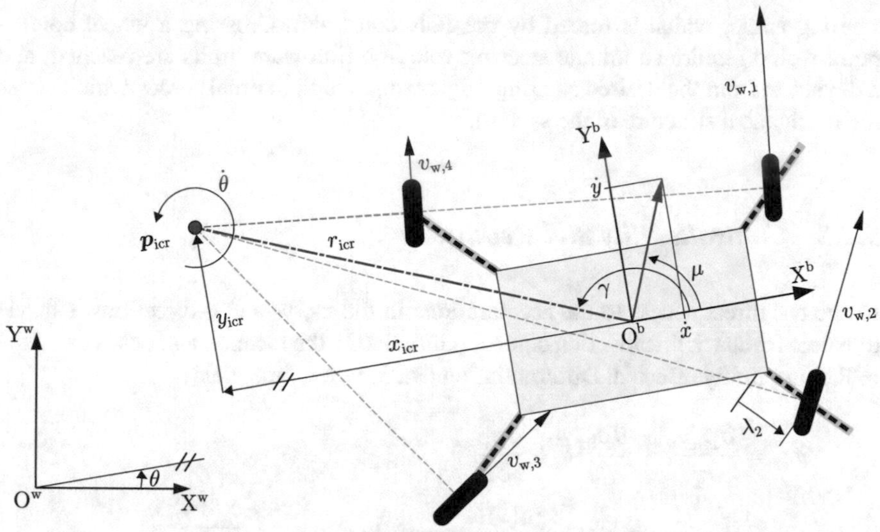

Fig. 3.7 The wheels align to the instantaneous center of rotation, which is located at $\boldsymbol{p}_{icr}$. The rotation between body frame (superscript b) and world frame (superscript w) is θ. The leg lengths λ_i (for $i = 1 \ldots 4$) are constant

The angle μ in Fig. 3.7 determines the direction of motion in the body frame. An offset of $\pm\pi/2$ to γ exists, where the sign depends on the direction of rotation of the platform. The spinning velocities of the four wheels are expressed by $v_{w,1}$ to $v_{w,4}$. A consistent motion requires each wheel to orient such that its direction of motion is perpendicular to the connection line to the ICR, see Fig. 3.7. This assumption is limited to the case of rigid body motions. This is not the case if a leg length λ_i is varying, i.e. $\dot{\lambda}_i \neq 0$. However, the only difference is that the wheel gets an additional velocity component in leg motion direction. The velocity component resulting from the ICR constraint remains unaffected.

It follows from (3.25) and (3.27) that the mapping $(x_{icr}, y_{icr}) \rightarrow (\dot{x}, \dot{y}, \dot{\theta})$ is not unique. That is, several motion states of the platform lead to the same ICR. For example, doubling v and $\dot{\theta}$ simultaneously does not shift $\boldsymbol{p}_{icr}$. As a consequence, just setting a specific location of the ICR is not sufficient to control the whole system uniquely. Furthermore, it can be seen that a mathematical singularity arises when the angular velocity $\dot{\theta}$ approaches zero, see (3.25). In terms of practical interpretation, this case implies that the radius of curvature tends to infinity inducing a pure translational motion. Hence, these mathematical problems have to be dealt with when controlling a wheeled mobile robot solely via the ICR [CPHV08]. In the approach considered here, a dynamic feedback linearization is used for kinematic control of the mobile base [GFASH09]. Hence, the explicit use of the ICR representation (3.25) is restricted to the problematic regions around the wheel steering axes.

Nevertheless, another type of singularity might emerge during motion. If the ICR passes the wheel very closely, the steering rate increases rapidly to follow the desired steering angle, which is forced by the ICR constraint. Crossing a wheel contact point would require an infinite steering velocity. Hardware limits are reached, and a deviation from the desired steering angle causes high internal forces which stress the mechanical structure of the system.

3.2.2 Controllability and Repulsion

If one has direct access to the accelerations in the x-, y- and θ-directions, e.g. via dynamic feedback linearization control [GFASH09], the location and behavior of the ICR is implicitly affected. Differentiating (3.25) w. r. t. time yields

$$\dot{\boldsymbol{p}}_{\text{icr}} = \frac{\partial \boldsymbol{p}_{\text{icr}}}{\partial \gamma}\dot{\gamma} + \frac{\partial \boldsymbol{p}_{\text{icr}}}{\partial \dot{\theta}}\ddot{\theta} + \frac{\partial \boldsymbol{p}_{\text{icr}}}{\partial v}\dot{v}$$
$$= \underbrace{\begin{pmatrix} -\frac{v}{|\dot{\theta}|}\sin(\gamma) & -\frac{\text{sgn}(\dot{\theta})v}{\dot{\theta}^2}\cos(\gamma) \\ \frac{v}{|\dot{\theta}|}\cos(\gamma) & -\frac{\text{sgn}(\dot{\theta})v}{\dot{\theta}^2}\sin(\gamma) \end{pmatrix}}_{\boldsymbol{J}_{\text{sap}}} \begin{pmatrix} \dot{\gamma} \\ \ddot{\theta} \end{pmatrix} + \begin{pmatrix} \frac{\cos(\gamma)}{|\dot{\theta}|} \\ \frac{\sin(\gamma)}{|\dot{\theta}|} \end{pmatrix} \dot{v} \tag{3.28}$$

as the ICR velocity in the plane, where $\boldsymbol{J}_{\text{sap}} \in \mathbb{R}^{2\times 2}$ is a Jacobian matrix used for the singularity avoidance of the platform (sap). This velocity can be directly controlled via $\dot{\gamma}$ and $\ddot{\theta}$ as long as $\boldsymbol{J}_{\text{sap}}$ has full rank, i.e. $\det(\boldsymbol{J}_{\text{sap}}) \neq 0$. It results from

$$\det(\boldsymbol{J}_{\text{sap}}) = \frac{v^2}{\dot{\theta}^3} \tag{3.29}$$

that a loss of controllability occurs in case of a pure rotational motion ($\dot{\theta} \neq 0$, $v = 0$) or when the angular velocity $\dot{\theta}$ tends to $\pm\infty$ for $v \neq 0$. Both conditions imply that the ICR is lying in the platform center. There is no need to control the ICR in that zone, since no singular configuration of the platform is reached.

An intuitive interpretation of the controllability via $\dot{\gamma}$ and $\ddot{\theta}$ can be given when considering Fig. 3.7 and (3.25): A variation in $\dot{\gamma}$ makes the ICR turn around the platform center (tangentially) while $\ddot{\theta}$ induces a radial motion. In summary, the two variables $\dot{\gamma}$ and $\ddot{\theta}$ have orthogonal effect on the ICR behavior, which provides a proper control input to push away the ICR in any direction, if necessary. In the approach, the total translational platform velocity is kept constant ($\dot{v} = 0$). Hence, the second part in (3.28) vanishes. Of course, this is only one particular choice.

In the area where the repulsion is required, i.e. around the steering axes, a potential analogous to (3.8) can be applied. The repulsive force $F_{\text{sap}}(d_{\text{sap}})$ is a function of the distance between the location of the ICR and the nearest wheel $\boldsymbol{p}_{\text{wheel}}$ (in body coordinates):

Fig. 3.8 Position and shape of the potentials in the body frame

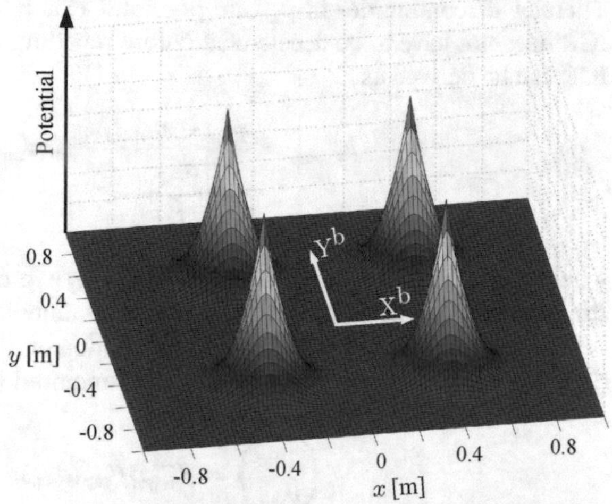

Fig. 3.9 Potential field extensions

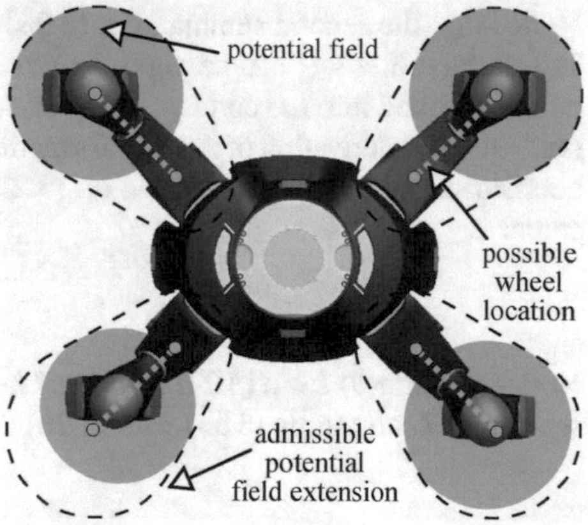

$$d_{\text{sap}} = \left\| \boldsymbol{p}_{\text{icr}} - \boldsymbol{p}_{\text{wheel}} \right\| . \tag{3.30}$$

Figure 3.8 illustrates the placement of the potential fields in the body frame for the case of fixed leg lengths. In case of leg motions, it is useful to restrict the potential field extension depending on the position of the wheel in order to avoid overlaps. Figure 3.9 illustrates such a relation between leg length and field extension.

3.2.3 Effect on the Instantaneous Center of Rotation

The force $F_{\text{sap}}(d_{\text{sap}})$ cannot be applied to the ICR directly, because the ICR does not possess an inertia. Thus the conversion via m/(Ns) is introduced. It has been shown in (3.28) that the platform accelerations have a direct effect on the ICR velocity.

Thereby, discontinuities in $\dot{\boldsymbol{p}}_{\text{icr}}$ are possible. This is in fact beneficial because the ICR does not have to be decelerated before repelling it. The repulsive effect on the ICR can be derived as

$$\dot{\boldsymbol{p}}_{\text{icr,sap}} = \underbrace{\frac{\boldsymbol{p}_{\text{icr}} - \boldsymbol{p}_{\text{wheel}}}{d_{\text{sap}}}}_{\boldsymbol{\eta}} F_{\text{sap}}(d_{\text{sap}}) \cdot \frac{\text{m}}{\text{Ns}}, \tag{3.31}$$

where $\dot{\boldsymbol{p}}_{\text{icr,sap}}$ is the required ICR velocity and $\boldsymbol{\eta}$ expresses the normalized direction from wheel contact point to ICR. Additional damping in (3.31) is not necessary because no kinetic energy is stored in the motion of the ICR. Combining (3.28) and (3.31) and setting $\dot{\boldsymbol{p}}_{\text{icr}} = \dot{\boldsymbol{p}}_{\text{icr,sap}}$ leads to the required values

$$\begin{pmatrix} \dot{\gamma}_{\text{sap}} \\ \ddot{\theta}_{\text{sap}} \end{pmatrix} = \boldsymbol{J}_{\text{sap}}^{-1} \boldsymbol{\eta} F_{\text{sap}}(d_{\text{sap}}) \cdot \frac{\text{m}}{\text{Ns}}. \tag{3.32}$$

Notice that the second summand in (3.28) is omitted due to the constraint $\dot{v} = 0$. The acceleration $\ddot{\theta}_{\text{sap}}$ can be applied by the motion controller, whereas $\dot{\gamma}_{\text{sap}}$ has to be transformed into the corresponding translational accelerations $\ddot{x}_{\text{sap}}$ and $\ddot{y}_{\text{sap}}$. The mathematical derivation of the transformation is straightforward and based on simple geometrical considerations. Based on (3.27), the ICR velocity around the platform center is

$$\dot{\gamma} = \frac{\dot{x}\ddot{y} - \dot{y}\ddot{x}}{\dot{x}^2 + \dot{y}^2}. \tag{3.33}$$

Notice that $\text{sgn}(\dot{\theta})$ from (3.27) is omitted here because $\dot{\theta} \neq 0$ holds in case of active repulsion. Rearranging (3.33) with (3.26) delivers the instantaneous linear relation

$$\ddot{y} = \frac{\dot{y}}{\dot{x}}\ddot{x} + \frac{v^2}{\dot{x}}\dot{\gamma} \tag{3.34}$$

between $\ddot{y}$ and $\ddot{x}$ for satisfying an arbitrary velocity $\dot{\gamma}$. To keep the translational velocity v constant, the time derivative of the velocity vector and the vector itself must be orthogonal, solely allowing a vector rotation. The condition is met by

$$\ddot{y} = -\frac{\dot{x}}{\dot{y}}\ddot{x}. \tag{3.35}$$

Applying the required value $\dot{\gamma}_{\text{sap}}$, the solution of the linear system of equations (3.34) and (3.35) is

$$\begin{pmatrix} \ddot{x}_{\text{sap}} \\ \ddot{y}_{\text{sap}} \end{pmatrix} = \dot{\gamma}_{\text{sap}} \begin{pmatrix} -\dot{y} \\ \dot{x} \end{pmatrix}. \tag{3.36}$$

A kinematic controller such as the dynamic feedback linearization [GFASH09] is able to realize the accelerations $\ddot{x}_{\text{sap}}$, $\ddot{y}_{\text{sap}}$, and $\ddot{\theta}_{\text{sap}}$.

3.2.4 Effect on the Wheel

In vehicles with variable footprint, it is also possible to repel the wheel from the ICR. In contrast to the control of the ICR (Sect. 3.2.3), a leg motion does not result in a deviation from the nominal trajectory in the plane. To apply a repulsive force to the wheel, an admittance-based mass-spring-damper equation

$$m_{\mathrm{w}}\ddot{\lambda}_{i,\mathrm{sap}} + d_{\mathrm{w}}\dot{\lambda}_{i,\mathrm{sap}} + k_{\mathrm{w}}\left(\lambda_{i,\mathrm{sap}} - \lambda_{i,0}\right) = F_{\mathrm{w}} \tag{3.37}$$

can be used, where m_{w}, d_{w}, and k_{w} describe the virtual mass, damper, and spring of wheel i. The parameter $\lambda_{i,0}$ describes the user-defined equilibrium position for the wheel location along the leg direction, and F_{w} denotes the control input resulting from the repulsive force to be applied. The reason for imposing the feedback gains d_{w} and k_{w} is the limitation of the leg length. For high performance around the equilibrium configuration, a nonlinear, increasing stiffness $k_{\mathrm{w}} = k_{\mathrm{w}}\left(\lambda_{i,\mathrm{sap}}\right)$ can be used, e.g.

$$k_{\mathrm{w}}\left(\lambda_{i,\mathrm{sap}}\right) = c_1\left(\lambda_{i,\mathrm{sap}} - \lambda_{i,0}\right)^2 \tag{3.38}$$

with $c_1 \in \mathbb{R}^+$. The repulsive force $F_{\mathrm{sap}}(d_{\mathrm{sap}})$ from Sect. 3.2.3 has to be mapped via

$$F_{\mathrm{w}} = -\boldsymbol{\eta}^T \boldsymbol{l} F_{\mathrm{sap}}(d_{\mathrm{sap}}), \tag{3.39}$$

wherein the first part describes the projection in the normalized fixed-leg direction $\boldsymbol{l}$ of the nearest wheel $\boldsymbol{p}_{\mathrm{wheel}}$. By inserting (3.38) and (3.39) into (3.37), an expression for the required leg acceleration can be developed:

$$\ddot{\lambda}_{i,\mathrm{sap}} = -\frac{\boldsymbol{\eta}^T \boldsymbol{l} F_{\mathrm{sap}}(d_{\mathrm{sap}})}{m_{\mathrm{w}}} - \frac{d_{\mathrm{w}}}{m_{\mathrm{w}}}\dot{\lambda}_{i,\mathrm{sap}} - \frac{c_1}{m_{\mathrm{w}}}\left(\lambda_{i,\mathrm{sap}} - \lambda_{i,0}\right)^3. \tag{3.40}$$

3.2.5 Control Design

So far, seven singularity avoidance accelerations have been determined: $\ddot{x}_{\mathrm{sap}}$, $\ddot{y}_{\mathrm{sap}}$, $\ddot{\theta}_{\mathrm{sap}}$, and $\ddot{\lambda}_{i,\mathrm{sap}}$ for $i = 1\ldots4$. If directly applied, a deviation from the nominal trajectory of the platform results. An additional control loop can be employed to lead back to the nominal trajectory as soon as the ICR has left the potential field. Figure 3.10 depicts the approach. The feedback gains $k_1 = k_1(d_{\mathrm{sap}})$ and $k_2 = k_2(d_{\mathrm{sap}})$ are active once the potential field is left. Their values are dependent on the desired closed-loop poles of the second-order system. The platform singularity avoidance acceleration is $\ddot{\boldsymbol{r}}_{\mathrm{sap}} = [\ddot{x}_{\mathrm{sap}}\ \ddot{y}_{\mathrm{sap}}\ \ddot{\theta}_{\mathrm{sap}}\ \ddot{\lambda}_{1,\mathrm{sap}}\ \ddot{\lambda}_{2,\mathrm{sap}}\ \ddot{\lambda}_{3,\mathrm{sap}}\ \ddot{\lambda}_{4,\mathrm{sap}}]^T$.

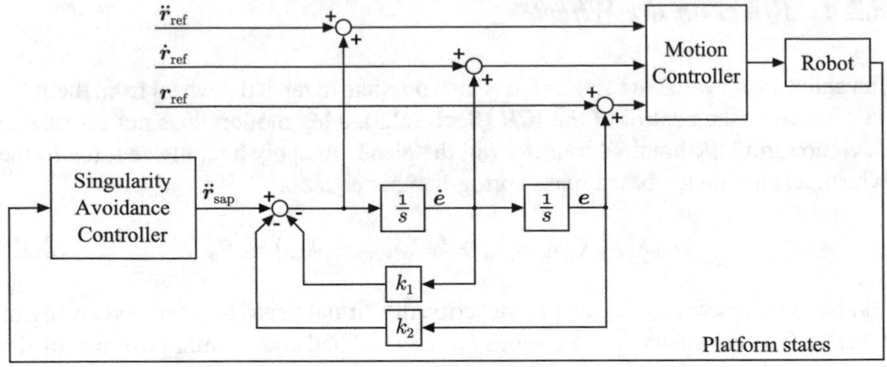

Fig. 3.10 Control loop to eliminate deviations induced by the platform singularity avoidance ($\boldsymbol{r}_{ref}, \dot{\boldsymbol{r}}_{ref}, \ddot{\boldsymbol{r}}_{ref}$: nominal trajectory, $\ddot{\boldsymbol{r}}_{sap}$: singularity avoidance acceleration, $\boldsymbol{e}$: position error, $\dot{\boldsymbol{e}}$: velocity error)

3.2.6 Simulations and Experiments

The control algorithm has been validated for the mobile platform kinematics of Rollin' Justin in simulation and has been implemented on the real system. The trajectory (in x, y and θ) consists of quintic Bézier splines, the platform achieves a rotational speed of 2.1 rad/s, which requires the maximum spinning rates of the wheels (13.9 rad/s). Initially, the leg lengths are kept constant.

In Fig. 3.11 (top), the ICR location is plotted within the time range (t_0, t_1). The potential fields (shaded circles) are crossed in case of deactivated control (solid) and avoided while the repulsion is active (dashed and dotted). The corresponding steering velocities are provided in Fig. 3.11 (bottom). The steering rates stay reasonably small while ICR control is active, but they display high peaks in the inactive case. Critical situations occur at $t = 9$ s, $t = 15$ s and $t = 18$ s. At this point it shall be mentioned that the deviation induced by the ICR control may lead to a different behavior after leaving a potential field. At $t = 9$ s the ICR is repelled from wheel 3 (lower left corner in Fig. 3.11, top). Before coming back onto the nominal trajectory, the ICR approaches wheel 1 (upper right corner), while the path of the ICR does not cross the potential field if the avoidance is deactivated. Therefore, the steering rate of wheel 1 increases slightly at $t = 10.5$ s.

Figure 3.12 shows the commanded accelerations. The bottom diagram indicates whether ICR repulsion is active or not (measurement case). The position errors stay within a range of ± 7 cm (translation) and $\pm 15°$ (rotation) during the whole motion. They can be reduced by using weaker potential fields. A trade-off between low steering rate and high tracking performance has to be found.

Activating the additional leg DOF reduces the required accelerations in translational and rotational direction because repelling a wheel from the ICR alleviates the need for the ICR to be pushed away. To support that idea, a simulated leg motion maneuver is shown in Fig. 3.13 (top). Starting position for the wheel contact point is

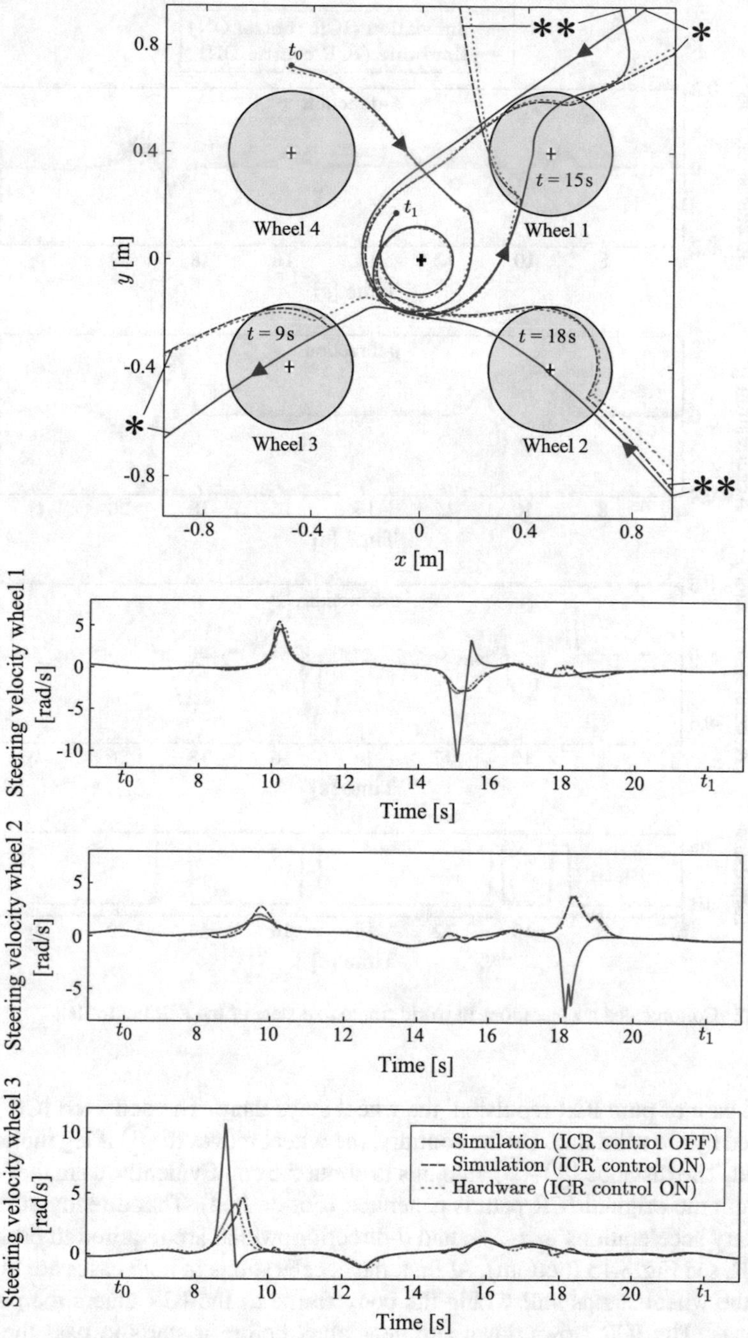

Fig. 3.11 ICR location in body frame and corresponding steering velocities of the wheels. The artificial potential fields of three wheels are penetrated

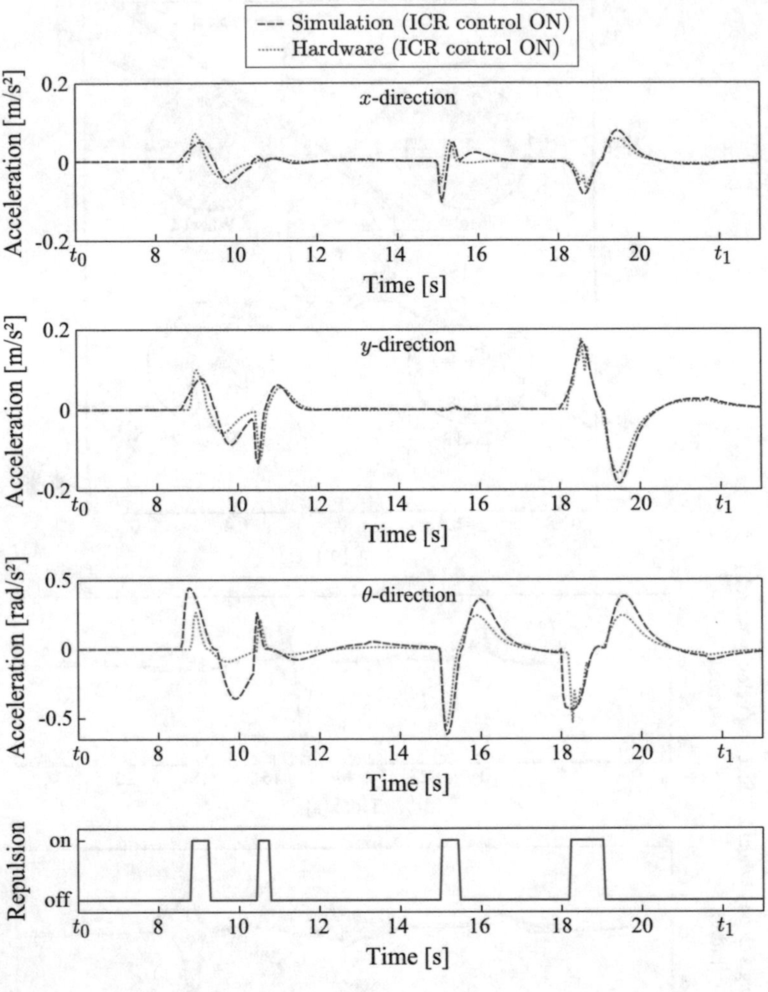

Fig. 3.12 Commanded accelerations in world frame and state of the ICR controller

Ⓐ. In case of pure ICR repulsion, the wheel stays there. The deflected ICR path is depicted by the solid line. On the contrary, the wheel moves to Ⓑ if leg motions are enabled. The distance Ⓐ–Ⓑ amounts to about 5.5 cm. Evidently, a smaller deflection from the original ICR path is generated (dot-dashed). That directly affects the necessary accelerations in x-, y-, and θ-direction, which are required to push away the ICR, see Fig. 3.13 (bottom). At first, the accelerations in both cases are identical since the wheel stands still within the body frame as the ICR enters the potential field at t_2. The ICR slows down and gets stuck before it starts to pass the wheel clockwise. Meanwhile, the wheel starts to move towards Ⓑ in case of activated leg motions. Eventually, that reduces the effort to push away the ICR and results in the less deformed path. The leg deviation plot (bottom/right) shows the simulated

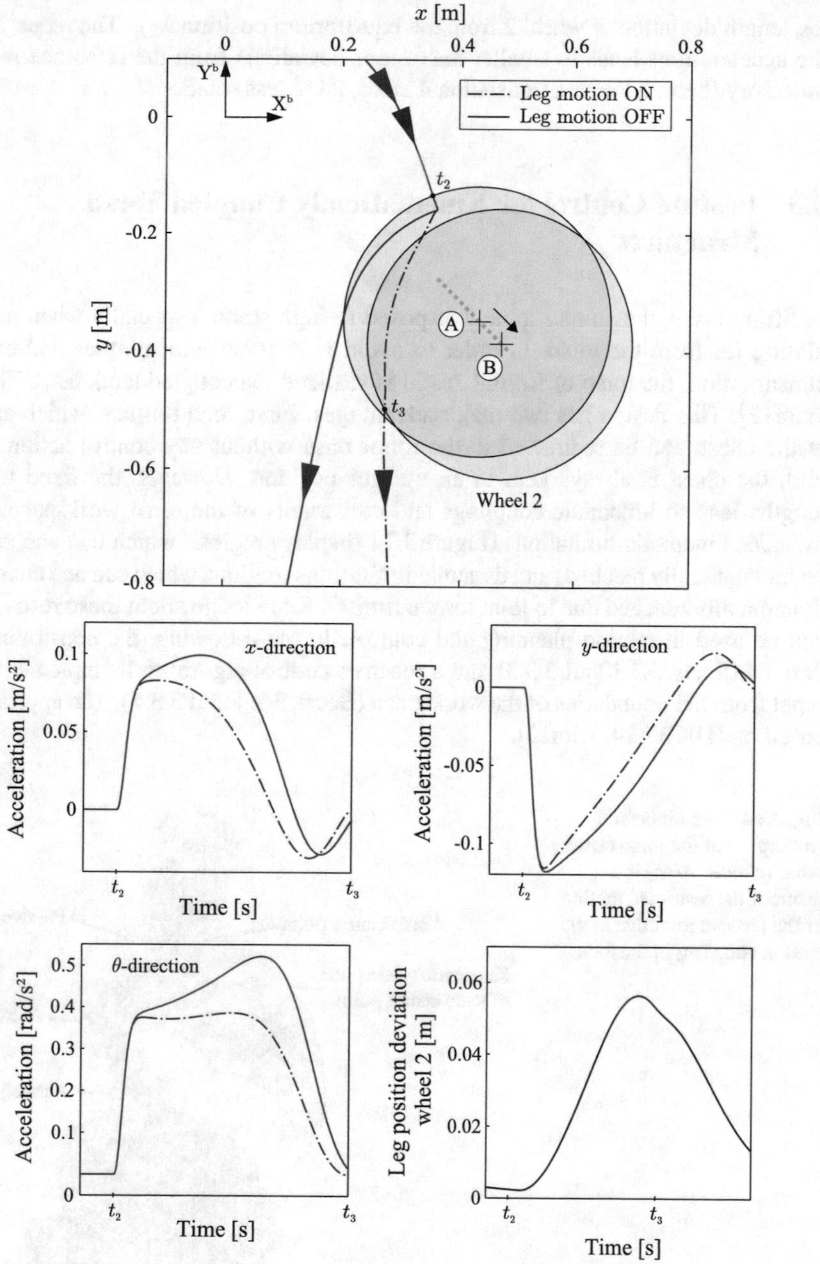

Fig. 3.13 ICR plot in body frame (*top*): Comparison between singularity avoidance with and without leg motions for the second wheel. In case of activated leg motions, the ICR repulsion is smaller (*bottom*), which reduces the position error

leg length deviation of wheel 2 from the equilibrium position $\lambda_{2,0}$. The reduction of the accelerations leads to smaller maximum deviations from the reference motion trajectory (here: 32 % less translational error, 24 % less rotational error).

3.3 Posture Control for Kinematically Coupled Torso Structures

In lifting tasks, the human spine is exposed to high strain, especially when manipulating far from the torso. In order to avoid high torso joint torques and energy consumption, the torso of Rollin' Justin is realized via coupled tendons [OEF+06, Wim12]. This design has two major advantages: First, load torques, which appear at the chest, can be redirected to the robot base without any control action. Second, the chest is always kept in an upright position. However, the fixed tendon lengths lead to kinematic couplings and constraints of the torso workspace. That includes kinematic limitations (Figure 3.14 displays regions which can and cannot be kinematically reached) and dynamic limitations (regions which can and cannot be dynamically reached due to joint torque limits). Knowledge about these restrictions can be used in motion planning and control. In the following, the constraints are derived (Sects. 3.3.1 and 3.3.3) and a reactive control algorithm is implemented to repel from the boundaries of the workspace (Sects. 3.3.4 and 3.3.5). The approach is based on [DKW+14, Kim13].

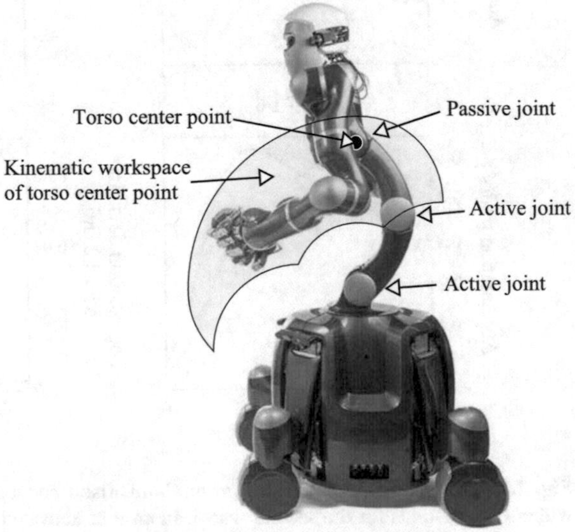

Fig. 3.14 The kinematic workspace of the torso center point (*shaded area*) is a result of the restricted motion of the passive joint due to the tendon coupling in the torso

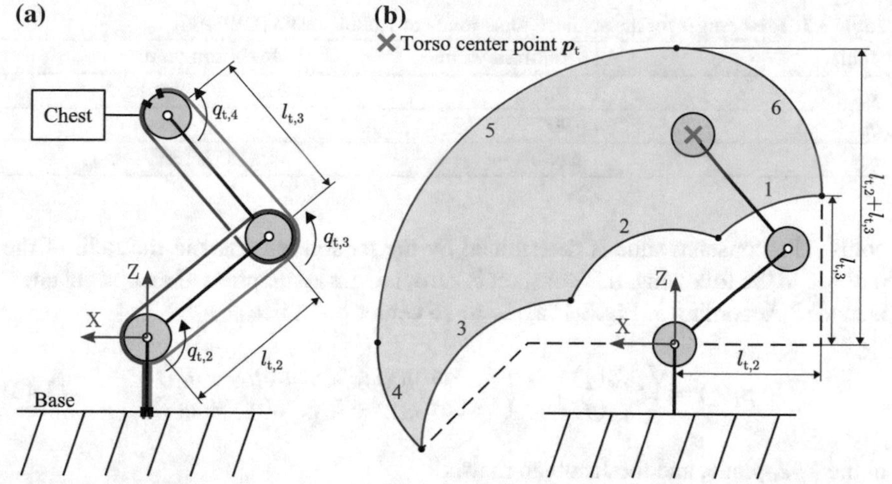

Fig. 3.15 Tendon routing and workspace of the torso of Rollin' Justin. The first torso joint (*vertical axis*) is not considered here. **a** Sketch of the torso illustrating the tendon kinematics. This is a modified figure taken from [Wim12]. **b** The workspace of the torso center point (*shaded area*) is bounded by the kinematics of the tendons. This is a modified figure taken from [Kim13]

3.3.1 Model of the Torso of Rollin' Justin

The torso of Rollin' Justin contains one passive and three actuated joints, cf. Sect. 2.3. Tendons are used to kinematically couple the torso joints two, three, and four. Therefore, only these joints $q_{t,2}$, $q_{t,3}$, and $q_{t,4}$ are considered in the following (t for "torso").[5] Figure 3.15a depicts the tendon routing. Motors are placed at $q_{t,2}$ and $q_{t,3}$. The upper joint $q_{t,4}$ is passive and ensures that the chest is always kept in an upright configuration. Of course, elasticities in the tendons exist, so that the kinematic constraint of an upright torso is an assumption that only holds for infinite tendon stiffnesses. However, experimental evaluations revealed that the joint angle deflections are very small [Rau13]. More detailed information on the mechanics of the torso can be found in [OEF+06, Wim12].

3.3.2 Kinematic Constraints

Assuming that the tendons are inelastic, the kinematic joint constraint

$$q_{t,2} + q_{t,3} + q_{t,4} = \text{const.} \tag{3.41}$$

[5]The vertical axis with joint value $q_{t,1}$ can be treated separately because it is not kinematically coupled.

Table 3.2 Joint ranges for the actuated torso joints of Rollin' Justin [OEF+06]

Joint	Minimum value	Maximum value
$q_{t,1}$	$-140°$	$200°$
$q_{t,2}$	$-90°$	$90°$
$q_{t,3}$	$\max(0°, -q_{t,2})$	$\min(135°, 135° - q_{t,2})$

holds. The constant value is determined by the tendon lengths and the radii of the pulleys. In the following, this constant is zero, i.e. the assumption of an upright chest is made.[6] According to Fig. 3.15a, the torso center point lies at

$$\boldsymbol{p}_{\mathrm{t}}(\boldsymbol{q}) = \begin{pmatrix} p_{\mathrm{t},x}(\boldsymbol{q}) \\ p_{\mathrm{t},z}(\boldsymbol{q}) \end{pmatrix} = \begin{pmatrix} l_{\mathrm{t},2}\sin(q_{\mathrm{t},2}) + l_{\mathrm{t},3}\sin(q_{\mathrm{t},2}+q_{\mathrm{t},3}) \\ l_{\mathrm{t},2}\cos(q_{\mathrm{t},2}) + l_{\mathrm{t},3}\cos(q_{\mathrm{t},2}+q_{\mathrm{t},3}) \end{pmatrix} \tag{3.42}$$

in the X–Z-plane, and the Jacobian matrix

$$\begin{aligned} \boldsymbol{J}_{\mathrm{t}}(\boldsymbol{q}) &= \frac{\partial \boldsymbol{p}_{\mathrm{t}}(\boldsymbol{q})}{\partial (q_{\mathrm{t},2}, q_{\mathrm{t},3})^T} \\ &= \begin{pmatrix} l_{\mathrm{t},2}\cos(q_{\mathrm{t},2}) + l_{\mathrm{t},3}\cos(q_{\mathrm{t},2}+q_{\mathrm{t},3}) & l_{\mathrm{t},3}\cos(q_{\mathrm{t},2}+q_{\mathrm{t},3}) \\ -l_{\mathrm{t},2}\sin(q_{\mathrm{t},2}) - l_{\mathrm{t},3}\sin(q_{\mathrm{t},2}+q_{\mathrm{t},3}) & -l_{\mathrm{t},3}\sin(q_{\mathrm{t},2}+q_{\mathrm{t},3}) \end{pmatrix} \end{aligned} \tag{3.43}$$

relates the joint velocities to the Cartesian velocities. In Table 3.2, the mechanical ranges of the joints are listed. Along with (3.42), one yields the kinematic workspace of the torso center point (3.15b), cf. [OEF+06]. More details on the mathematical derivations and the analytical expressions of the workspace boundaries are provided in Appendix A.

3.3.3 Dynamic Constraints

Depending on the configuration of the torso, the load

$$\boldsymbol{F}_{\mathrm{t}} = \begin{pmatrix} F_{\mathrm{t},x} \\ F_{\mathrm{t},z} \end{pmatrix} = F_{\mathrm{t,load}} \begin{pmatrix} \cos(\alpha) \\ -\sin(\alpha) \end{pmatrix}, \tag{3.44}$$

which acts in the X–Z-plane at the torso center point, requires specific motor torques to counteract:

$$\begin{aligned} \tau_{\mathrm{t}} &= -\boldsymbol{J}_{\mathrm{t}}(\boldsymbol{q})^T \boldsymbol{F}_{\mathrm{t}} \\ &= -F_{\mathrm{t,load}} \begin{pmatrix} l_{\mathrm{t},3}\cos(q_{\mathrm{t},2}+q_{\mathrm{t},3}-\alpha) + l_{\mathrm{t},2}\cos(q_{\mathrm{t},2}-\alpha) \\ l_{\mathrm{t},3}\cos(q_{\mathrm{t},2}+q_{\mathrm{t},3}-\alpha) \end{pmatrix}. \end{aligned} \tag{3.45}$$

[6] In the zero configuration, all torso links are aligned with the vertical axis.

The magnitude of the force is given by $F_{\mathrm{t,load}}$ and its direction is defined by the angle α about the Y-axis. Using the range of the feasible motor torques ("min" for minimum, "max" for maximum)

$$\tau_{\mathrm{t,2,min}} \leq \tau_{\mathrm{t,2}} \leq \tau_{\mathrm{t,2,max}}, \tag{3.46}$$

$$\tau_{\mathrm{t,3,min}} \leq \tau_{\mathrm{t,3}} \leq \tau_{\mathrm{t,3,max}}, \tag{3.47}$$

additional workspace boundaries can be derived to comply with the actuator limits. These restrictions are called dynamic constraints because the load force $\boldsymbol{F}_{\mathrm{t}}$ can be directly linked to the dynamic equations of the torso, i.e. the constraints incorporate gravitational effects (weight of the upper body, tools), inertial effects, external forces (contact between upper body and the environment), and Coriolis/centrifugal effects.

The boundary concerning $\tau_{\mathrm{t,2}}$ can be obtained from the first line of (3.45) by inserting the forward kinematics (3.42). The result

$$\frac{\tau_{\mathrm{t,2}}}{F_{\mathrm{t,load}}} = -x \sin(\alpha) - z \cos(\alpha) \tag{3.48}$$

represents straight lines in the X–Z-plane. Inserting the boundaries $\tau_{\mathrm{t,2,min}}$ and $\tau_{\mathrm{t,2,max}}$ from (3.46) delivers the new, dynamic workspace constraints. Their geometric interpretation is very intuitive: The ratios $\tau_{\mathrm{t,2,min}}/F_{\mathrm{t,load}}$ and $\tau_{\mathrm{t,2,max}}/F_{\mathrm{t,load}}$ determine the distance between the boundaries while α describes the decline of the belt for feasible torques in the second torso joint, see Fig. 3.16.

The boundary concerning $\tau_{\mathrm{t,3}}$ can be calculated by solving the second line of (3.45) for $q_{\mathrm{t,2}} + q_{\mathrm{t,3}}$, which delivers

$$q_{\mathrm{t,2}} + q_{\mathrm{t,3}} = \alpha \pm \arccos\left(\frac{-\tau_{\mathrm{t,3}}}{F_{\mathrm{t,load}} l_{\mathrm{t,3}}}\right). \tag{3.49}$$

Once again, one can apply the limits on the motor torque, i.e. $\tau_{\mathrm{t,3,min}}$ and $\tau_{\mathrm{t,3,max}}$ from (3.47). If $\tau_{\mathrm{t,3}}$, α, and $F_{\mathrm{t,load}}$ are constant, $q_{\mathrm{t,2}}+q_{\mathrm{t,3}}$ is obviously constant too. Moreover, (3.49) states that there only exists a configuration if $-1 \leq -\tau_{\mathrm{t,3}}/(F_{\mathrm{t,load}} l_{\mathrm{t,3}}) \leq 1$. If this inequality is not fulfilled, no constraints on the workspace are imposed by $\tau_{\mathrm{t,3}}$. That effect can be interpreted as the minimum/maximum torque being large enough to counteract the given load force in any part of the workspace [Kim13]. Analogous to (3.48), one can express the workspace boundaries in the X–Z-representation (Fig. 3.15) by using the forward kinematics (3.42):

$$\left(x - l_{\mathrm{t,3}} \sin(q_{\mathrm{t,2}} + q_{\mathrm{t,3}})\right)^2 + \left(z - l_{\mathrm{t,3}} \cos(q_{\mathrm{t,2}} + q_{\mathrm{t,3}})\right)^2 = l_{\mathrm{t,2}}^2. \tag{3.50}$$

Hence, the general constraint is geometrically described by a circle with radius $l_{\mathrm{t,2}}$ and its center at

$$x = l_{\mathrm{t,3}} \sin\left(\alpha \pm \arccos\left(\frac{-\tau_{\mathrm{t,3}}}{F_{\mathrm{t,load}} l_{\mathrm{t,3}}}\right)\right), \tag{3.51}$$

$$z = l_{t,3} \cos\left(\alpha \pm \arccos\left(\frac{-\tau_{t,3}}{F_{t,load} l_{t,3}}\right)\right). \tag{3.52}$$

Due to the cases in (3.49) and the torque limits $\tau_{t,3,min}$ and $\tau_{t,3,max}$, four boundaries of type (3.51) and (3.52) exist. Details (center locations, workspace specifications) are provided in Appendix A. The four circle centers span a rectangle. According to (3.51) and (3.52), their locations depend on the load as well as the torque limits. In case of $\tau_{t,2,min} = \tau_{t,3,min} = -\tau_{t,2,max} = -\tau_{t,3,max}$, the centers even lie on the straight $\tau_{t,2,min}$ and $\tau_{t,2,max}$ boundaries. That can easily be proven by inserting (3.51) and (3.52) into (3.48). Figure 3.16 illustrates the overall workspace resulting from intersecting the kinematic and the dynamic workspace of the torso. The four scenarios depict the workspace for different loads with different effective directions for the case of ±230 Nm minimum/maximum torque for each actuated joint. The

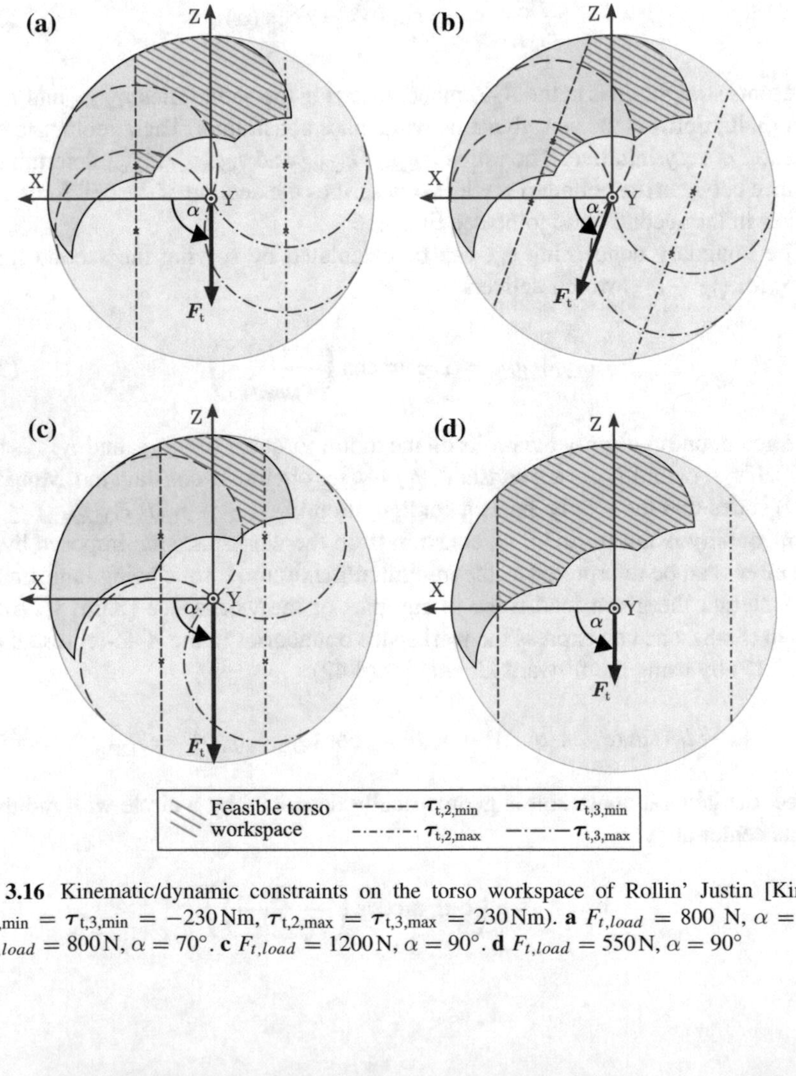

Fig. 3.16 Kinematic/dynamic constraints on the torso workspace of Rollin' Justin [Kim13] ($\tau_{t,2,min} = \tau_{t,3,min} = -230$ Nm, $\tau_{t,2,max} = \tau_{t,3,max} = 230$ Nm). **a** $F_{t,load} = 800$ N, $\alpha = 90°$. **b** $F_{t,load} = 800$ N, $\alpha = 70°$. **c** $F_{t,load} = 1200$ N, $\alpha = 90°$. **d** $F_{t,load} = 550$ N, $\alpha = 90°$

torque constraints on torso joint two are parallel to the direction of $\boldsymbol{F}_{\mathrm{t}}$, cf. (3.48). The dynamic constraints for joint three are actually defined by semicircles instead of full circles. This is due to the fact that $q_{\mathrm{t},3} \geq 0$ always holds according to Table 3.2. Only one half of each circle ensures a constant angle sum $q_{\mathrm{t},2} + q_{\mathrm{t},3}$ that complies with (3.41). Invalid workspace areas exist between two circular dynamic constraint boundaries resulting from the same minimum torque (or maximum torque, respectively) of the joint. There, the joint torque would exceed the maximum torque or go below the minimum torque. If $F_{\mathrm{t,load}}$ gets smaller, the semicircles approach each other (Fig. 3.16c → Fig. 3.16a), hence the workspace grows. At the point of overlapping, the constraints vanish, see Fig. 3.16d. The load specification in Fig. 3.16d with $F_{\mathrm{t,load}} = 550\,\mathrm{N}$ and $\alpha = 90°$ represents a typical (static) lifting task of an object of about 10 kg.

3.3.4 Control Design

A repulsive potential $V_{\mathrm{tws}}(\boldsymbol{q})$ related to the torso workspace (tws) can be applied to repel the torso from its kinematic and dynamic boundaries:

$$\boldsymbol{\tau}_{\mathrm{tws}} = -\left(\frac{\partial V_{\mathrm{tws}}(\boldsymbol{q})}{\partial \boldsymbol{q}}\right)^T - \boldsymbol{D}_{\mathrm{tws}}(\boldsymbol{q})\dot{\boldsymbol{q}} + \boldsymbol{g}(\boldsymbol{q}). \tag{3.53}$$

Kinetic energy dissipation is implemented by the positive definite damping matrix $\boldsymbol{D}_{\mathrm{tws}}(\boldsymbol{q})$ analogous to Sect. 3.1.3.2, where the Hessian matrix $\partial^2 V_{\mathrm{tws}}(\boldsymbol{q})/\partial \boldsymbol{q}^2$ of the potential and the reflected mass in constraint direction are utilized. More details on the control design can be found in [DKW+14].

3.3.5 Experiments

This experiment on Rollin' Justin shows how the controller forces the torso to stay in the admissible workspace. The user interacts with the robot in gravity compensation mode and moves the torso in the X–Z-plane. The path of the torso center point is illustrated in Fig. 3.17 (left). The damping ratio is set to $\xi = 0.3$, the load force is specified by $F_{\mathrm{t,load}} = 470\,\mathrm{N}$ and $\alpha = 90°$. The minimum and maximum torques are set to ±140 Nm, the maximum stiffness in the potential is 1000 N/m, and the starting distance is $d_0 = 0.1\,\mathrm{m}$. An additional singularity avoidance is applied to avoid the outstretched torso configuration. The repulsive and damping forces along the path, the repulsion state of the controller, and the relevant joint torques are depicted in Fig. 3.17 (right). Although the torso center point stays in the admissible workspace, the torque in joint three violates the dynamic constraints between 40 and 50 s by about 6 Nm. This is due to the fact that inertial and Coriolis/centrifugal effects are not taken into account in the control law. Sufficiently large safety margins have been used, i.e. ±140 Nm as minimum/maximum torque instead of the actual limit of ±230 Nm, in

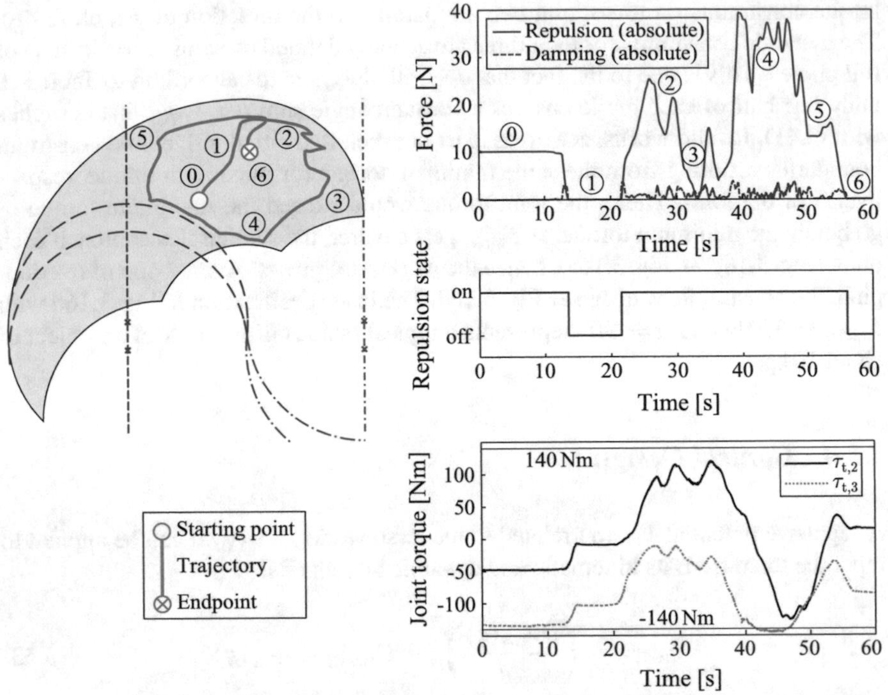

Fig. 3.17 The torso is manually moved in gravity compensation mode in the X–Z-plane without consideration of any constraints by the user himself. The instants ⓪ to ⑥ in the path plot (*left*) correspond to the ones in the measurements (*right*)

order to circumvent the need to compute a dynamic model online while providing a safe and simple mechanism.

3.4 Classical Objectives in Reactive Control

A wide variety of different task descriptions based on potential fields has been applied to robotic systems, and the number is steadily increasing. Three state-of-the-art methods are recapitulated briefly in the following, since they are employed in the whole-body control framework in Chap. 6.

3.4.1 Cartesian Impedance

A controller in the Cartesian coordinates of the end-effector is a powerful tool to perform various tasks. Since the trajectories and the contact behavior can be defined in the Cartesian space of the tool itself, this method is ideally suited for end-effector tasks within a whole-body control framework. While the idea of impedance control has already been developed decades ago [Hog85], the availability of modern force-torque-controlled robots has given a new impetus to research in Cartesian impedance [OASKH08, ASOH07]. The control torque in the general case is given by

$$\boldsymbol{\tau}_{\text{Cart}} = -\left(\frac{\partial V_{\text{Cart}}(\boldsymbol{x}(\boldsymbol{q}), \boldsymbol{x}_{\text{des}}(t))}{\partial \boldsymbol{q}}\right)^T - \boldsymbol{D}_{\text{Cart}}(\boldsymbol{q})\dot{\boldsymbol{q}}, \tag{3.54}$$

where the artificial potential $V_{\text{Cart}}(\boldsymbol{x}(\boldsymbol{q}), \boldsymbol{x}_{\text{des}}(t))$ is defined in the Cartesian operational space coordinates $\boldsymbol{x}(\boldsymbol{q})$ of the end-effector, realizing the trajectory $\boldsymbol{x}_{\text{des}}(t)$. Additional damping can be implemented via the positive definite matrix $\boldsymbol{D}_{\text{Cart}}(\boldsymbol{q})$.

3.4.2 Manipulator Singularity Avoidance

The kinematic manipulability measure [Yos90]

$$m_{\text{kin}}(\boldsymbol{q}) = \sqrt{\det(\boldsymbol{J}(\boldsymbol{q})\boldsymbol{J}(\boldsymbol{q})^T)} = \sigma_1 \cdot \sigma_2 \cdot \ldots \cdot \sigma_m \tag{3.55}$$

w.r.t. the Jacobian matrix $\boldsymbol{J}(\boldsymbol{q}) \in \mathbb{R}^{m \times n}$ for $m < n$ describes the kinematic ability to move the end-effector in its m directions ($m = 6$ for full Cartesian control). If one or more singular values of the Jacobian matrix, denoted by σ_1 to σ_m, approach zero, the manipulability goes to zero as well. The feasible motion of the end-effector gets restricted in the singularity. Based on (3.55), one can design a potential to keep the measure at a reasonably high value for manipulator singularity avoidance (msa) [Ott08]:

$$V_{\text{msa}}(\boldsymbol{q}) = \begin{cases} k_{\text{msa}}(m_{\text{kin}}(\boldsymbol{q}) - m_{\text{kin},0})^2 & \text{for } m_{\text{kin}}(\boldsymbol{q}) \leq m_{\text{kin},0}(\boldsymbol{q}) \\ 0 & \text{for } m_{\text{kin}}(\boldsymbol{q}) > m_{\text{kin},0}(\boldsymbol{q}) \end{cases}. \tag{3.56}$$

The artificial potential is non-zero when $m_{\text{kin}}(\boldsymbol{q})$ is smaller than a specified threshold $m_{\text{kin},0}(\boldsymbol{q}) \in \mathbb{R}^+$. One can parameterize the controller through the gain $k_{\text{msa}} \in \mathbb{R}^+$. The control torque is

$$\boldsymbol{\tau}_{\text{msa}} = -\left(\frac{\partial V_{\text{msa}}(\boldsymbol{q})}{\partial \boldsymbol{q}}\right)^T - \boldsymbol{D}_{\text{msa}}(\boldsymbol{q})\dot{\boldsymbol{q}}, \tag{3.57}$$

where additional damping can be implemented via the positive definite matrix $\boldsymbol{D}_{\text{msa}}(\boldsymbol{q})$. The controller structure of (3.57) with (3.56) is taken from [Ott08].

3.4.3 Avoidance of Mechanical End Stops

Repulsive potentials to avoid the mechanical end stops of joints are applied to robotic systems frequently [MCR96, SK05]. Consider a mechanically feasible joint range given by $q_{i,\min} \leq q_i \leq q_{i,\max}$ for joint i. To avoid reaching the minimum or maximum joint values $q_{i,\min}$ and $q_{i,\max}$, a repulsive potential can be set up:

$$V_{i,\mathrm{mes}}(q_i) = \begin{cases} k_{i,\mathrm{mes}}(q_i - q_{i,\mathrm{lower}})^2 & \text{for } q_i \leq q_{i,\mathrm{lower}} \\ 0 & \text{for } q_{i,\mathrm{lower}} < q_i < q_{i,\mathrm{upper}} \\ k_{i,\mathrm{mes}}(q_i - q_{i,\mathrm{upper}})^2 & \text{for } q_i \geq q_{i,\mathrm{upper}} \end{cases} \tag{3.58}$$

The stiffness is determined by $k_{i,\mathrm{mes}} \in \mathbb{R}^+$ and the potential starts at $q_{i,\mathrm{lower}}$ and $q_{i,\mathrm{upper}}$, respectively. The control action is calculated following

$$\boldsymbol{\tau}_{\mathrm{mes}} = -\left(\frac{\partial\left(\sum_{i=1}^{n} V_{i,\mathrm{mes}}(q_i)\right)}{\partial \boldsymbol{q}}\right)^T - \boldsymbol{D}_{\mathrm{mes}}(\boldsymbol{q})\dot{\boldsymbol{q}}, \tag{3.59}$$

where damping can be applied by means of the positive definite matrix $\boldsymbol{D}_{\mathrm{mes}}(\boldsymbol{q})$. Usually, one would specify a diagonal shape for $\boldsymbol{D}_{\mathrm{mes}}(\boldsymbol{q})$ since the joints are not coupled, except for mechanical designs like the kinematically coupled torso of Rollin' Justin as described in Sect. 3.3.

3.5 Summary

In this Chapter 3, several reactive methods have been developed based on artificial potential fields. The definition of tasks through attractive and repulsive potentials is advantageous thanks to the simple and intuitive controller parameterization. The controller gains have a direct relation to the physical world, e.g. the potential stiffness which can be interpreted as the stiffness of a virtual spring.

The goal of the task in Sect. 3.1 was to avoid self-collisions in a reactive way. By means of a geometric collision model, the distances between potentially colliding parts of the body were computed, and these links were repelled from each other. Additionally, a configuration-dependent damping has been introduced to assign damping ratios in the collision directions to dissipate kinetic energy and prevent oscillations. The second reactive task in Sect. 3.2 extended the kinematic control of wheeled platforms by a singularity avoidance. When the instantaneous center of rotation approaches the steering axis of a wheel, then the required steering velocity for consistent locomotion goes to infinity. The proposed algorithm repels the instantaneous center of rotation from the axis to avoid this critical situation. In Sect. 3.3, repulsive forces were generated to fully exploit the kinematic and dynamic workspaces of the tendon-coupled torso of Rollin' Justin. All reactive controllers in this chapter have been validated in experiments.

The developed methods extend the set of classical tasks in robotics. A characteristic of the controllers in this Chapter 3 is that they do not access all actuated joints of the robot at the same time. Therefore, they are particularly suitable to be combined in a multi-objective whole-body control framework, where the tasks are realized simultaneously. The outcomes of this Chapter 3 will be directly applied within the hierarchical redundancy resolution treated in the next chapter.

Chapter 4
Redundancy Resolution by Null Space Projections

Robots with many DOF and several simultaneous objectives necessarily require a redundancy resolution. In most state-of-the-art approaches, one solves optimization problems for a hierarchical arrangement of the involved tasks. The highest-priority task is executed employing all capabilities of the robotic system. The second-priority task is then performed in the null space of this highest-priority task. In other words, the task on the second level is executed as well as possible without disturbing the first level. The task on level three is then executed without disturbing the two higher-priority tasks, and so forth.

The literature distinguishes two basic approaches: An optimization problem is formulated and solved via dedicated solvers such as in [DSBDS09, KLW11, EMW14]. An advantage of these concepts is that both equality and inequality constraints can be integrated in the task hierarchy. However, the numerical costs of the methods strongly depend on the used solver and the optimization problem formulation. The second kind of approach is more frequently used and solves the optimization problem by means of pseudoinverses [DMB93] and projection matrices. These so-called *null space projections* have been mainly developed in the 1980s [Lie77, Kha87, NHY87, SS91]. Today they are standard tools in kinematic control [BB04, NCM+08, AIC09, SGJG10, LMP11] and dynamic control [ASOFH03, KSPW04, SK05, NCM+08, MKK09, SVKS13]. In torque-controlled robots, a control input is processed by the null space projector related to all higher-priority tasks and the resulting torque then executes the desired task as well as possible without interfering with the higher-level objectives. However, the structure of these null space projectors only allows direct implementation of equality constraints. Nevertheless, even inequality constraints can be integrated by modifying the projectors as will be shown later.

In the Sects. 4.1–4.3, different null space projections are investigated and compared in simulations and experiments [DOAS15]. The property of strictness is treated in Sect. 4.1 by analyzing the successive and the augmented null space projection. In Sect. 4.2, three different types of projection consistencies are compared, namely the

A. Dietrich, *Whole-Body Impedance Control of Wheeled Humanoid Robots*,
Springer Tracts in Advanced Robotics 116, DOI 10.1007/978-3-319-40557-5_4

statically consistent one, the dynamically consistent one, and the new concept of stiffness-consistent projections. The direct comparison in simulations and experiments is performed in Sect. 4.3. In addition to the extensive survey of different techniques, these sections contribute to a deeper understanding of dynamic consistency in general, i.e. when the redundancy resolution dynamically decouples the priority levels. Moreover, it is shown that on hardware, the theoretically inferior, statically consistent null space projections feature an equal performance as the dynamically consistent resolutions. Since statically consistent null space projections do not require a (numerically expensive) estimate of the joint inertia matrix, these findings are of high relevance for robotics. In Sect. 4.4, the null space projectors are modified and enhanced to deal with inequality (or unilateral) constraints, singularities, and dynamic task hierarchies. Experiments on Rollin' Justin confirm the benefits of the approach.

4.1 Strictness of the Hierarchy

Consider a manipulator with n DOF and r tasks, which are defined by

$$\boldsymbol{x}_i = \boldsymbol{f}_i(\boldsymbol{q}) \in \mathbb{R}^{m_i} \; \forall i, \; 1 \leq i \leq r. \tag{4.1}$$

The dimension of the ith task is $m_i \leq n$. The differential mappings from joint velocities to task velocities are determined by the Jacobian matrices $\boldsymbol{J}_i(\boldsymbol{q}) \in \mathbb{R}^{m_i \times n}$ via

$$\dot{\boldsymbol{x}}_i = \boldsymbol{J}_i(\boldsymbol{q})\dot{\boldsymbol{q}} \; , \quad \boldsymbol{J}_i(\boldsymbol{q}) = \frac{\partial \boldsymbol{f}_i(\boldsymbol{q})}{\partial \boldsymbol{q}} . \tag{4.2}$$

In the following, $\boldsymbol{J}_i(\boldsymbol{q})$ is assumed to be non-singular, hence of full row rank.[1] The main task ($i = 1$) has dimension $m_1 < n$ such that a kinematic redundancy of $n - m_1$ DOF remains to execute subtasks in its null space. The hierarchy is defined such that $i = 1$ is top priority and $i_a < i_b$ implies that i_a has higher priority than i_b.

4.1.1 Successive Projections

A task torque $\boldsymbol{\tau}_2 \in \mathbb{R}^n$ on the second priority level can be projected onto the null space of the main task via

$$\boldsymbol{\tau}_2^{\mathrm{p}} = \boldsymbol{N}_2^{\mathrm{suc}}(\boldsymbol{q})\boldsymbol{\tau}_2, \tag{4.3}$$

[1] Dealing with singular matrices or changing rank requires additional treatment, both in kinematic and torque control [DW95, Chi97, DASH12, DWASH12a]. This aspect will be addressed in Sect. 4.4.

where $\boldsymbol{\tau}_2^{\mathrm{p}} \in \mathbb{R}^n$ is the projected torque that does not interfere with the main task. The successive null space projector $\boldsymbol{N}_2^{\mathrm{suc}}(\boldsymbol{q})$ is obtained by evaluating

$$\boldsymbol{N}_2^{\mathrm{suc}}(\boldsymbol{q}) = \boldsymbol{I} - \boldsymbol{J}_1(\boldsymbol{q})^T (\boldsymbol{J}_1(\boldsymbol{q})^{\#})^T, \tag{4.4}$$

wherein $\{\}^{\#}$ represents the generalized inverse[2] and $\boldsymbol{I}$ is the identity matrix. Analogous to (4.3), the subtasks in the hierarchy $(2 < i \leq r)$ can be implemented by

$$\boldsymbol{\tau}_i^{\mathrm{p}} = \boldsymbol{N}_i^{\mathrm{suc}}(\boldsymbol{q})\boldsymbol{\tau}_i \tag{4.5}$$

with the null space projectors obtained via the *successive* formula

$$\boldsymbol{N}_i^{\mathrm{suc}}(\boldsymbol{q}) = \boldsymbol{N}_{i-1}^{\mathrm{suc}}(\boldsymbol{q}) \left(\boldsymbol{I} - \boldsymbol{J}_{i-1}(\boldsymbol{q})^T (\boldsymbol{J}_{i-1}(\boldsymbol{q})^{\#})^T\right). \tag{4.6}$$

One obtains the control torque by summing the main task torque and all projected torques

$$\boldsymbol{\tau} = \boldsymbol{\tau}_1 + \sum_{i=2}^{r} \boldsymbol{\tau}_i^{\mathrm{p}}. \tag{4.7}$$

This technique has been analyzed for inverse kinematics in [Ant09]. Note that $\boldsymbol{N}_i^{\mathrm{suc}}(\boldsymbol{q})$ is strictly speaking not a mathematical projector because it is not idempotent in general. However, in robotics the term null space projector is commonly used in this context though, since (4.6) is based on the fundamental idea of null space projections, and the single matrix (4.4) for the second level is idempotent.

4.1.2 Augmented Projections

The augmented approach [SS91] is identical to the successive projection on the first null space level (4.3), (4.4). From the third level on, the projected torque is determined via

$$\boldsymbol{\tau}_i^{\mathrm{p}} = \boldsymbol{N}_i^{\mathrm{aug}}(\boldsymbol{q})\boldsymbol{\tau}_i, \tag{4.8}$$

where the null space projector $\boldsymbol{N}_i^{\mathrm{aug}}(\boldsymbol{q})$ is given by

$$\boldsymbol{N}_i^{\mathrm{aug}}(\boldsymbol{q}) = \boldsymbol{I} - \boldsymbol{J}_{i-1}^{\mathrm{aug}}(\boldsymbol{q})^T (\boldsymbol{J}_{i-1}^{\mathrm{aug}}(\boldsymbol{q})^{\#})^T. \tag{4.9}$$

The *augmented* Jacobian matrix $\boldsymbol{J}_{i-1}^{\mathrm{aug}}(\boldsymbol{q})$ contains all higher-priority Jacobian matrices:

---

[2] Since $\{\}^{\#}$ is not unique, the particular choice for the inverse has an influence on the projected torques. That aspect will be addressed in Sect. 4.2.

$$J_{i-1}^{\text{aug}}(q) = \begin{pmatrix} J_1(q) \\ J_2(q) \\ \vdots \\ J_{i-1}(q) \end{pmatrix}. \tag{4.10}$$

The control torque is obtained via (4.7) again by using (4.8) instead of (4.5). The direct implementation of (4.9) is computationally expensive due to the large number of rows in $J_{i-1}^{\text{aug}}(q)$ and the resulting complexity in the pseudoinversion. Usually, recursive algorithms [SS91, BB98, SK05] are applied to reduce the numerical effort:

$$N_1^{\text{aug}} = I, \tag{4.11}$$

$$\hat{J}_{i-1}(q) = J_{i-1}(q) N_{i-1}^{\text{aug}}(q)^T, \tag{4.12}$$

$$N_i^{\text{aug}}(q) = N_{i-1}^{\text{aug}}(q) \left(I - \hat{J}_{i-1}(q)^T (\hat{J}_{i-1}(q)^{\#})^T \right) \tag{4.13}$$

for $2 \leq i \leq r$. The term $\hat{J}_i(q) \in \mathbb{R}^{m_i \times n}$ describes the Jacobian matrix of level i projected onto the null space of all higher-priority tasks. In fact, the additional intermediate step (4.12) is the only difference between the successive and the augmented approach.

4.2 Consistency of the Projections

While Sect. 4.1 investigated the overall structure of the hierarchy, the projector *consistency* determines how the null space itself is defined in terms of properties and shape. Prior to that analysis, the pseudoinverse of a matrix is briefly introduced [DMB93]. In Sect. 4.1.1 the generalized inverse $\{\}^{\#}$ was applied but it was not specified. A generalized inverse $A^{\#}$ of a full-row-rank matrix $A \in \mathbb{R}^{m \times n}$ with $m < n$ has to satisfy the criterion

$$AA^{\#} = I \tag{4.14}$$

of right inverses. One can find an infinite number of generalized inverses that meet (4.14). As of now, the notation $\{\}^{W+}$ is used instead of $\{\}^{\#}$ to disambiguate the inverse by the weighting matrix $W \in \mathbb{R}^{n \times n}$. One can formulate

$$A^{W+} = W^{-1} A^T \left(A W^{-1} A^T \right)^{-1}, \tag{4.15}$$

which fulfills (4.14) as long as the inversion on the right is feasible.[3] The use of such generalized inverses is very common in robotics, especially in inverse kinematics. In the following, the effects of the weighting matrix are clarified and classified into three different types of torque control projection consistencies. The analysis is

[3]The term $AW^{-1}A^T$ has to be of rank m, and W must be invertible.

performed on a two-level system for the sake of simplicity, yet all statements can be transferred to more complex hierarchies without loss of generality. The distinction between successive and augmented projection does not have to be made here since $\boldsymbol{N}_2(\boldsymbol{q}) = \boldsymbol{N}_2^{\text{suc}}(\boldsymbol{q}) = \boldsymbol{N}_2^{\text{aug}}(\boldsymbol{q})$. The dynamic equations (2.6) of a robot with n DOF are used in the following.

4.2.1 Static Consistency

Definition 4.1 *A null space projector $\boldsymbol{N}_j(\boldsymbol{q}) \in \mathbb{R}^{n\times n}$ is said to be "statically consistent" if a subtask does not generate interfering forces in the operational spaces of all higher-priority tasks in any static equilibrium. The condition*

$$\boldsymbol{N}_j(\boldsymbol{q})\boldsymbol{J}_i(\boldsymbol{q})^T = \boldsymbol{0} \tag{4.16}$$

for $i < j$ must hold in any steady state with $\dot{\boldsymbol{q}} = \ddot{\boldsymbol{q}} = \boldsymbol{0}$.

To explain Definition 4.1, the following example can be considered. The control action

$$\boldsymbol{\tau} = \boldsymbol{g}(\boldsymbol{q}) + \boldsymbol{J}_1(\boldsymbol{q})^T\boldsymbol{F}_1 + \boldsymbol{N}_2(\boldsymbol{q})\boldsymbol{\tau}_2 \tag{4.17}$$

is applied, which includes a gravity compensation, the execution of a main task via the operational space force $\boldsymbol{F}_1 \in \mathbb{R}^{m_1}$, and a null space action $\boldsymbol{\tau}_2$ processed by the null space projector $\boldsymbol{N}_2(\boldsymbol{q})$. A static equilibrium is considered, i.e. the control action (4.17) does not generate any motions. Inserting (4.17) into the quasi-static version of (2.6) with $\dot{\boldsymbol{q}} = \ddot{\boldsymbol{q}} = \boldsymbol{0}$ and reorganizing the terms yields

$$-\boldsymbol{\tau}_{\text{ext}} = \boldsymbol{J}_1(\boldsymbol{q})^T\boldsymbol{F}_1 + \boldsymbol{N}_2(\boldsymbol{q})\boldsymbol{\tau}_2, \tag{4.18}$$

where $\boldsymbol{\tau}_2$ can be decomposed into

$$\boldsymbol{\tau}_2 = \boldsymbol{J}_1(\boldsymbol{q})^T\boldsymbol{F}_{2,\boldsymbol{J}_1} + \mathcal{N}(\boldsymbol{J}_1(\boldsymbol{q}))^T\boldsymbol{F}_{2,\mathcal{N}(\boldsymbol{J}_1)}. \tag{4.19}$$

The condition $\boldsymbol{J}_1(\boldsymbol{q})\mathcal{N}(\boldsymbol{J}_1(\boldsymbol{q}))^T = \boldsymbol{0}$ is fulfilled, where $\mathcal{N}(\boldsymbol{J}_1(\boldsymbol{q})) \in \mathbb{R}^{(n-m_1)\times n}$ describes the directions orthogonal to $\boldsymbol{J}_1(\boldsymbol{q})$. Thus $\boldsymbol{F}_{2,\boldsymbol{J}_1} \in \mathbb{R}^{m_1}$ is the contribution of the null space action $\boldsymbol{\tau}_2$ in the main task space, and $\boldsymbol{F}_{2,\mathcal{N}(\boldsymbol{J}_1)} \in \mathbb{R}^{n-m_1}$ refers to the remaining space. Inserting (4.19) into (4.18) yields

$$-\boldsymbol{\tau}_{\text{ext}} = \boldsymbol{J}_1(\boldsymbol{q})^T\boldsymbol{F}_1 + \underbrace{\boldsymbol{N}_2(\boldsymbol{q})\boldsymbol{J}_1(\boldsymbol{q})^T}_{=\boldsymbol{0}}\boldsymbol{F}_{2,\boldsymbol{J}_1} + \boldsymbol{N}_2(\boldsymbol{q})\mathcal{N}(\boldsymbol{J}_1(\boldsymbol{q}))^T\boldsymbol{F}_{2,\mathcal{N}(\boldsymbol{J}_1)}. \tag{4.20}$$

Due to $\boldsymbol{N}_2(\boldsymbol{q})\boldsymbol{J}_1(\boldsymbol{q})^T = \left(\boldsymbol{I} - \boldsymbol{J}_1(\boldsymbol{q})^T(\boldsymbol{J}_1(\boldsymbol{q})^{W+})^T\right)\boldsymbol{J}_1(\boldsymbol{q})^T = \boldsymbol{0}$, the force $\boldsymbol{F}_{2,\boldsymbol{J}_1}$ has no influence in (4.20) [ASOFH03] as described in Definition 4.1. This result can also be interpreted as the confirmation of the considered equilibrium. Although

$\boldsymbol{\tau}_2 \neq \mathbf{0}$ in general, the contribution of the null space action in main task direction is filtered.

One choice for the weighting matrix is

$$\boldsymbol{W} = \boldsymbol{I}, \tag{4.21}$$

so that for $\boldsymbol{A} = \boldsymbol{J}_1(\boldsymbol{q})$ one can write (4.15) as

$$\boldsymbol{J}_1(\boldsymbol{q})^{\boldsymbol{I}+} = \boldsymbol{J}_1(\boldsymbol{q})^+ = \boldsymbol{J}_1(\boldsymbol{q})^T \left(\boldsymbol{J}_1(\boldsymbol{q})\boldsymbol{J}_1(\boldsymbol{q})^T\right)^{-1}. \tag{4.22}$$

In the notation of this so-called *Moore-Penrose pseudoinverse*, the identity in the superscript is often omitted. Compared to other weighting matrices, this choice is computationally very cheap. Moreover, the null space can be interpreted from a geometric point of view [DWASH12b], and damped least-squares techniques can be applied easily [DW95].

4.2.2 Dynamic Consistency

The property of static consistency is shared by all null space projectors, independent of the weighting matrix. But apart from static consistency, specific weighting matrices have *additional* advantages such as the so-called dynamic consistency treated in this section. The main difference is that static consistency only guarantees that the hierarchy levels do not interfere in a steady state, while dynamic consistency guarantees that they also do not interfere during the transient into this steady state.

Definition 4.2 *A null space projector $\boldsymbol{N}_j(\boldsymbol{q}) \in \mathbb{R}^{n\times n}$ is said to be "dynamically consistent" [Kha95] if a subtask does not generate accelerations in the operational spaces of all higher-priority tasks. The condition*

$$\boldsymbol{J}_i(\boldsymbol{q})\boldsymbol{M}(\boldsymbol{q})^{-1}\boldsymbol{N}_j(\boldsymbol{q}) = \mathbf{0} \tag{4.23}$$

for $i < j$ must always hold.

The operational space dynamics based on (2.6) can be written as

$$\ddot{\boldsymbol{x}}_1 = -\boldsymbol{J}_1(\boldsymbol{q})\boldsymbol{M}(\boldsymbol{q})^{-1}\left(\boldsymbol{C}(\boldsymbol{q},\dot{\boldsymbol{q}})\dot{\boldsymbol{q}} + \boldsymbol{g}(\boldsymbol{q}) - \boldsymbol{\tau}_{\text{ext}}\right) + \dot{\boldsymbol{J}}_1(\boldsymbol{q},\dot{\boldsymbol{q}})\dot{\boldsymbol{q}} + \boldsymbol{J}_1(\boldsymbol{q})\boldsymbol{M}(\boldsymbol{q})^{-1}\boldsymbol{\tau} \tag{4.24}$$

after transformation in the main task directions $\boldsymbol{x}_1$.[4] For the sake of simplicity the term

$$\boldsymbol{p}_1(\boldsymbol{q},\dot{\boldsymbol{q}},\boldsymbol{\tau}_{\text{ext}}) = -\boldsymbol{J}_1(\boldsymbol{q})\boldsymbol{M}(\boldsymbol{q})^{-1}\left(\boldsymbol{C}(\boldsymbol{q},\dot{\boldsymbol{q}})\dot{\boldsymbol{q}} + \boldsymbol{g}(\boldsymbol{q}) - \boldsymbol{\tau}_{\text{ext}}\right) + \dot{\boldsymbol{J}}_1(\boldsymbol{q},\dot{\boldsymbol{q}})\dot{\boldsymbol{q}} \tag{4.25}$$

[4] Notice that $\ddot{\boldsymbol{x}}_i = \dot{\boldsymbol{J}}_i(\boldsymbol{q},\dot{\boldsymbol{q}})\dot{\boldsymbol{q}} + \boldsymbol{J}_i(\boldsymbol{q})\ddot{\boldsymbol{q}}$ due to the dependencies in the Jacobian matrices (4.2).

is introduced. The implementation of the control input

$$\boldsymbol{\tau} = \boldsymbol{J}_1(\boldsymbol{q})^T \boldsymbol{F}_1 + \boldsymbol{N}_2(\boldsymbol{q})\boldsymbol{\tau}_2 \tag{4.26}$$

with the main task force $\boldsymbol{F}_1 \in \mathbb{R}^{m_1}$ modifies (4.24) to

$$\ddot{\boldsymbol{x}}_1 = \boldsymbol{p}_1(\boldsymbol{q}, \dot{\boldsymbol{q}}, \boldsymbol{\tau}_{\text{ext}}) + \boldsymbol{\Lambda}_1(\boldsymbol{q})^{-1}\boldsymbol{F}_1 + \boldsymbol{J}_1(\boldsymbol{q})\boldsymbol{M}(\boldsymbol{q})^{-1}\boldsymbol{N}_2(\boldsymbol{q})\boldsymbol{\tau}_2, \tag{4.27}$$

where the reflected main task inertia $\boldsymbol{\Lambda}_1(\boldsymbol{q}) \in \mathbb{R}^{m_1 \times m_1}$ is defined as

$$\boldsymbol{\Lambda}_1(\boldsymbol{q}) = \left(\boldsymbol{J}_1(\boldsymbol{q})\boldsymbol{M}(\boldsymbol{q})^{-1}\boldsymbol{J}_1(\boldsymbol{q})^T\right)^{-1}. \tag{4.28}$$

The direct effect of the subtask torque $\boldsymbol{\tau}_2 \in \mathbb{R}^n$ on the main task acceleration $\ddot{\boldsymbol{x}}_1$ is determined by the coefficient of $\boldsymbol{\tau}_2$, i.e. (4.23) must be fulfilled for $i = 1$ and $j = 2$ to eliminate any effects of the lower-priority task on the main task acceleration. An intuitive interpretation of (4.23) is that the projector decouples the inertias on all priority levels. This beneficial decoupling property will be used in the stability analysis of Chap. 5.

4.2.2.1 Configuration-Dependent Weighting Matrix $W(q)$ Using the Inertia Matrix $M(q)$

Khatib [Kha87] has shown that the weighting matrix

$$\boldsymbol{W} = \boldsymbol{M}(\boldsymbol{q}) \tag{4.29}$$

fulfills (4.23) and the corresponding pseudoinverse minimizes the instantaneous kinetic energy of the manipulator. Another weighting matrix proposed by Park [Par99] is

$$\boldsymbol{W} = \boldsymbol{J}_1(\boldsymbol{q})^T\boldsymbol{J}_1(\boldsymbol{q}) + \boldsymbol{M}(\boldsymbol{q})\boldsymbol{Y}_1(\boldsymbol{q})^T\boldsymbol{Y}_1(\boldsymbol{q})\boldsymbol{M}(\boldsymbol{q}), \tag{4.30}$$

where $\boldsymbol{Y}_1(\boldsymbol{q}) \in \mathbb{R}^{(n-m_1)\times n}$ is a matrix which spans the null space of $\boldsymbol{J}_1(\boldsymbol{q})$. In fact, an infinite number of configuration-dependent weighting matrices $\boldsymbol{W}(\boldsymbol{q})$ exist that feature dynamic consistency. For a general formulation, the Jacobian matrix $\boldsymbol{J}_1(\boldsymbol{q})$ is decomposed via singular value decomposition [MK89] such that

$$\boldsymbol{J}_1(\boldsymbol{q}) = \boldsymbol{U}_1(\boldsymbol{q})\boldsymbol{S}_1(\boldsymbol{q})\boldsymbol{V}_1(\boldsymbol{q})^T, \tag{4.31}$$

where $\boldsymbol{U}_1(\boldsymbol{q}) \in \mathbb{R}^{m_1 \times m_1}$ and $\boldsymbol{V}_1(\boldsymbol{q}) \in \mathbb{R}^{n \times n}$ are orthonormal matrices, and $\boldsymbol{S}_1(\boldsymbol{q}) \in \mathbb{R}^{m_1 \times n}$ is a rectangular diagonal matrix containing the singular values σ_1 to σ_{m_1}. The null space can be geometrically interpreted when considering

$$\boldsymbol{V}_1(\boldsymbol{q}) = \left(\boldsymbol{X}_1(\boldsymbol{q})^T, \boldsymbol{Y}_1(\boldsymbol{q})^T\right). \tag{4.32}$$

The m_1 rows in $\boldsymbol{X}_1(\boldsymbol{q}) \in \mathbb{R}^{m_1 \times n}$ span the range space of $\boldsymbol{J}_1(\boldsymbol{q})$, while the $n - m_1$ rows in $\boldsymbol{Y}_1(\boldsymbol{q}) \in \mathbb{R}^{(n-m_1)\times n}$ span its null space. The orthogonality $\boldsymbol{X}_1(\boldsymbol{q})\boldsymbol{Y}_1(\boldsymbol{q})^T = \boldsymbol{0}$ holds. Inspired by (4.30) one can formulate a general rule for the weighting matrix $\boldsymbol{W}(\boldsymbol{q})$ that always fulfills the requirements of dynamic consistency:

$$\boldsymbol{W} = \boldsymbol{X}_1(\boldsymbol{q})^T \boldsymbol{X}_1(\boldsymbol{q}) \boldsymbol{B}_X + \boldsymbol{B}_Y \boldsymbol{Y}_1(\boldsymbol{q})^T \boldsymbol{Y}_1(\boldsymbol{q}) \boldsymbol{M}(\boldsymbol{q}). \tag{4.33}$$

The proof is provided in the Appendix B.2. Note that $\boldsymbol{W}(\boldsymbol{q})$ has to be non-singular to apply the pseudoinversion (4.15) where $\boldsymbol{W}(\boldsymbol{q})^{-1}$ is used. The matrices $\boldsymbol{B}_X, \boldsymbol{B}_Y \in \mathbb{R}^{n\times n}$ must fulfill

$$\operatorname{rank}(\boldsymbol{B}_X) \geq m_1, \tag{4.34}$$

$$\operatorname{rank}(\boldsymbol{B}_Y) \geq n - m_1. \tag{4.35}$$

These rank conditions are necessary but not sufficient to guarantee invertibility of $\boldsymbol{W}(\boldsymbol{q})$. Nevertheless, in Appendix B.2 it is shown that the condition on the rank of $\boldsymbol{B}_X$ can even be dropped when using another formulation than the one based on the pseudoinversion (4.15). With the knowledge of the general formulation, the weighting matrices of Khatib (4.29) and Park (4.30) can be regarded as special cases of (4.33) in fact.

$$\text{“Khatib” (4.29)}: \boldsymbol{B}_X = \boldsymbol{M}(\boldsymbol{q}),\ \boldsymbol{B}_Y = \boldsymbol{I}.$$

$$\text{“Park” (4.30)}: \boldsymbol{B}_X = \boldsymbol{J}_1(\boldsymbol{q})^T \boldsymbol{J}_1(\boldsymbol{q}),\ \boldsymbol{B}_Y = \boldsymbol{M}(\boldsymbol{q}).$$

Khatib [Kha87] found out that only one pseudoinverse satisfies (4.23). From that and the proof in Appendix B.2 one can conclude that any weighting matrix (4.33) leads to the identical pseudoinverse and null space projector. Thus a generalized inverse with (4.33) minimizes the instantaneous kinetic energy of the manipulator and the corresponding null space projector dynamically decouples the priority levels by block-diagonalizing the inertia matrix.

The respective null space projector has been shown to be load-independent [FK97]. Changing the load inertia on the higher-priority level does not result in a different null space projector indeed. When considering such an additional load or modified reflected inertia $\boldsymbol{L}_1 \in \mathbb{R}^{m_1\times m_1}$ and the altered joint inertia matrix

$$\boldsymbol{M}_{\oplus}(\boldsymbol{q}) = \boldsymbol{M}(\boldsymbol{q}) + \boldsymbol{J}_1(\boldsymbol{q})^T \boldsymbol{L}_1 \boldsymbol{J}_1(\boldsymbol{q}), \tag{4.36}$$

then the equality

$$\boldsymbol{N}_2(\boldsymbol{q}) = \boldsymbol{I} - \boldsymbol{J}_1(\boldsymbol{q})^T (\boldsymbol{J}_1(\boldsymbol{q})^{\boldsymbol{M}(\boldsymbol{q})+})^T = \boldsymbol{I} - \boldsymbol{J}_1(\boldsymbol{q})^T (\boldsymbol{J}_1(\boldsymbol{q})^{\boldsymbol{M}_{\oplus}(\boldsymbol{q})+})^T \tag{4.37}$$

holds. Load independence allows to ignore loads in the controller. Their estimation or measurement can be avoided and using such a null space projector decouples internal motions from load-dependent influences [FK97]. The invariance of the load can also be seen in the fact that (4.33) only requires the inertia matrix applied to the null space $\boldsymbol{Y}_1(\boldsymbol{q})$ and not necessarily to the range space $\boldsymbol{X}_1(\boldsymbol{q})$.

4.2.2.2 Arbitrary Weighting Matrix *W*

Another interesting type of dynamically consistent null space projector can be formulated which is acceleration-based originally:

$$\boldsymbol{N}_2(\boldsymbol{q}) = \boldsymbol{M}(\boldsymbol{q})\left(\boldsymbol{I} - \boldsymbol{J}_1(\boldsymbol{q})^{\boldsymbol{W}+}\boldsymbol{J}_1(\boldsymbol{q})\right)\boldsymbol{M}(\boldsymbol{q})^{-1}. \tag{4.38}$$

The premultiplication of $\boldsymbol{M}(\boldsymbol{q})$ is required for (4.23) and the multiplication by $\boldsymbol{M}(\boldsymbol{q})^{-1}$ from the right is necessary for the idempotence $\boldsymbol{N}_2(\boldsymbol{q}) = \boldsymbol{N}_2(\boldsymbol{q})\boldsymbol{N}_2(\boldsymbol{q})$. The major difference of (4.38) compared to the previous, dynamically consistent approach is that the null space projection is performed on acceleration level. If one considers $\boldsymbol{\tau}_2^{\mathrm{p}} = \boldsymbol{N}_2(\boldsymbol{q})\boldsymbol{\tau}_2$ with (4.38), the secondary task torque is initially transformed into a joint acceleration through the multiplication by $\boldsymbol{M}(\boldsymbol{q})^{-1}$. Then, a null space projection on acceleration level is applied as in standard kinematic robot control. Afterwards, this solution on acceleration level is transformed back into joint torques via $\boldsymbol{M}(\boldsymbol{q})$. The general idea of the procedure *torque→ acceleration → projected acceleration → projected torque* is intuitive and has been frequently implemented and analyzed [HS87, PMU+08]. The proof for dynamic consistency of (4.38) and the equivalence to the standard projector for the general weighting matrix (4.33) is provided in Appendix B.3.

One choice for the weighting matrix in (4.38) is $\boldsymbol{W} = \boldsymbol{I}$. Moreover, due to the standard Moore–Penrose pseudoinversion, singularity-robust techniques such as [DASH12] can be applied easier to preserve continuity in the control law. This projector can also be computed in a recursive way to reduce the numerical effort. The adaptation of (4.11)–(4.13) is

$$\boldsymbol{N}_1^{\mathrm{aug,s}} = \boldsymbol{I}, \tag{4.39}$$

$$\hat{\boldsymbol{J}}_{i-1}(\boldsymbol{q}) = \boldsymbol{J}_{i-1}(\boldsymbol{q})\boldsymbol{N}_{i-1}^{\mathrm{aug,s}}(\boldsymbol{q})^T, \tag{4.40}$$

$$\boldsymbol{N}_i^{\mathrm{aug,s}}(\boldsymbol{q}) = \boldsymbol{N}_{i-1}^{\mathrm{aug,s}}(\boldsymbol{q})\left(\boldsymbol{I} - \hat{\boldsymbol{J}}_{i-1}(\boldsymbol{q})^+\hat{\boldsymbol{J}}_{i-1}(\boldsymbol{q})\right) \tag{4.41}$$

$$\boldsymbol{N}_i^{\mathrm{aug}}(\boldsymbol{q}) = \boldsymbol{M}(\boldsymbol{q})\boldsymbol{N}_i^{\mathrm{aug,s}}(\boldsymbol{q})\boldsymbol{M}(\boldsymbol{q})^{-1}. \tag{4.42}$$

The matrices $\boldsymbol{N}_i^{\mathrm{aug,s}}(\boldsymbol{q}) \in \mathbb{R}^{n\times n}$ are auxiliary null space projectors on acceleration level, which are upgraded to dynamic consistency in (4.42). This solution has similar properties as the dynamically consistent solutions from before: Dynamic consistency and the idempotence criterion $\boldsymbol{N}_i^{\mathrm{aug}}(\boldsymbol{q}) = \boldsymbol{N}_i^{\mathrm{aug}}(\boldsymbol{q})\boldsymbol{N}_i^{\mathrm{aug}}(\boldsymbol{q})$ are fulfilled. However, load independence [FK97] is not featured. In the experiments conducted later, further differences will be demonstrated concerning the implementation on a real robot.

4.2.3 Stiffness Consistency

An increasing number of parallel elastic actuators (PEA) is encountered in the fields of prostheses, exoskeletons, and rehabilitation [DH08, WSA11, HTSG12, GEGS12]. Mounting mechanical springs in parallel to the motors allows to downsize the actuators because gravitational loads can be counterbalanced by the passive elements. Energy efficiency can be drastically improved that way, both from a static point of view (gravity compensation) and from a dynamic perspective (energy-efficient cyclic motions such as walking or jumping). The research group of Herr [AH09] has recently achieved impressive results in the field of active prostheses with additional passive elements where the principles of biomechanics and neural control are combined to design new devices.

Consider a scenario where a main task is statically accomplished by such a set of parallel mechanical springs, e.g. to keep the end-effector at a location by pre-adjusting the joints and springs such that no motor power is required to maintain the main task position and orientation. The so-called *stiffness-consistent* null space projector can then be used to simultaneously accomplish a secondary task while minimizing active regulation on the main task level by exploiting the springs. The dynamics (2.6) for constant external forces are extended by an additional joint spring $\boldsymbol{k}(\boldsymbol{q}, \boldsymbol{q}_0) \in \mathbb{R}^n$ such that

$$\boldsymbol{M}(\boldsymbol{q})\ddot{\boldsymbol{q}} + \boldsymbol{C}(\boldsymbol{q}, \dot{\boldsymbol{q}})\dot{\boldsymbol{q}} + \boldsymbol{g}(\boldsymbol{q}) + \boldsymbol{k}(\boldsymbol{q}, \boldsymbol{q}_0) = \boldsymbol{\tau} + \boldsymbol{\tau}_{\text{ext}} \tag{4.43}$$

holds, and $\boldsymbol{q}_0 \in \mathbb{R}^n$ is the equilibrium configuration where the spring counterbalances the gravitational load and the external forces.

Definition 4.3 *A null space projector $\boldsymbol{N}_j(\boldsymbol{q}_0) \in \mathbb{R}^{n\times n}$ is said to be "stiffness-consistent" if it is "statically consistent" and if a subtask does not cause static deviations in the operational spaces of all higher-priority tasks. These tasks with higher priority are partially or completely executed by springs $\boldsymbol{k}(\boldsymbol{q}, \boldsymbol{q}_0) \in \mathbb{R}^n$ with equilibrium configuration $\boldsymbol{q} = \boldsymbol{q}_0$. The condition*

$$\boldsymbol{J}_i(\boldsymbol{q}_0)\boldsymbol{K}(\boldsymbol{q}_0)^{-1}\boldsymbol{N}_j(\boldsymbol{q}_0) = \boldsymbol{0} \tag{4.44}$$

for $i < j$ must hold locally around the steady state $\boldsymbol{q} = \boldsymbol{q}_0$ with $\ddot{\boldsymbol{q}} = \dot{\boldsymbol{q}} = \boldsymbol{0}$, where the local stiffness matrix is

$$\boldsymbol{K}(\boldsymbol{q}_0) = \left.\frac{\partial \boldsymbol{k}(\boldsymbol{q}, \boldsymbol{q}_0)}{\partial \boldsymbol{q}}\right|_{\boldsymbol{q}=\boldsymbol{q}_0}. \tag{4.45}$$

In this equilibrium $\boldsymbol{q}_0$, the linearizations

$$\boldsymbol{k}_{\text{lin}}(\boldsymbol{q}, \boldsymbol{q}_0) = \boldsymbol{k}(\boldsymbol{q}_0) + \left.\frac{\partial \boldsymbol{k}(\boldsymbol{q}, \boldsymbol{q}_0)}{\partial \boldsymbol{q}}\right|_{\boldsymbol{q}=\boldsymbol{q}_0} (\boldsymbol{q} - \boldsymbol{q}_0) = \boldsymbol{k}(\boldsymbol{q}_0) + \boldsymbol{K}(\boldsymbol{q}_0)\Delta\boldsymbol{q}, \tag{4.46}$$

$$g_{\text{lin}}(\boldsymbol{q}, \boldsymbol{q}_0) = \boldsymbol{g}(\boldsymbol{q}_0) + \left.\frac{\partial \boldsymbol{g}(\boldsymbol{q})}{\partial \boldsymbol{q}}\right|_{\boldsymbol{q}=\boldsymbol{q}_0} (\boldsymbol{q} - \boldsymbol{q}_0) = \boldsymbol{g}(\boldsymbol{q}_0) + \boldsymbol{G}(\boldsymbol{q}_0)\Delta\boldsymbol{q} \tag{4.47}$$

can be evaluated where $\boldsymbol{K}(\boldsymbol{q}_0) \in \mathbb{R}^{n\times n}$ is the local, positive definite stiffness matrix in the equilibrium, $\boldsymbol{G}(\boldsymbol{q}_0) \in \mathbb{R}^{n\times n}$ describes the local, linear gravity behavior, and $\Delta\boldsymbol{q} = \boldsymbol{q} - \boldsymbol{q}_0$. At $\boldsymbol{q} = \boldsymbol{q}_0$, the counterbalance $\boldsymbol{k}(\boldsymbol{q}_0) = -\boldsymbol{g}(\boldsymbol{q}_0) + \boldsymbol{\tau}_{\text{ext}}$ holds for constant external forces. Then the quasi-static version of the dynamics (4.43) with

$$\boldsymbol{\tau} = \boldsymbol{N}_2(\boldsymbol{q}_0)\boldsymbol{\tau}_2 \tag{4.48}$$

yields

$$\boldsymbol{K}(\boldsymbol{q}_0)\Delta\boldsymbol{q} = -\boldsymbol{G}(\boldsymbol{q}_0)\Delta\boldsymbol{q} + \boldsymbol{N}_2(\boldsymbol{q}_0)\boldsymbol{\tau}_2. \tag{4.49}$$

Locally around the equilibrium the differential mapping (4.2) can be used to obtain

$$\Delta\boldsymbol{x}_1 = \boldsymbol{J}_1(\boldsymbol{q}_0)\boldsymbol{K}(\boldsymbol{q}_0)^{-1}\left(-\boldsymbol{G}(\boldsymbol{q}_0)\Delta\boldsymbol{q} + \boldsymbol{N}_2(\boldsymbol{q}_0)\boldsymbol{\tau}_2\right), \tag{4.50}$$

which has clear similarities to (4.27). If the weighting matrix

$$\boldsymbol{W} = \boldsymbol{K}(\boldsymbol{q}_0) \tag{4.51}$$

is chosen, the main task does not experience a direct disturbance by the lower-priority task control action $\boldsymbol{\tau}_2$. In other words, the contribution of the springs to the main task can be preserved this way. Note that a null space action usually leads to $\Delta\boldsymbol{q} \neq \boldsymbol{0}$, so that a small error will be indirectly generated due to the altered gravity torques. Nevertheless, the simulations in Sect. 4.3.1 will demonstrate that this effect is very limited. Any spring can be used for stiffness-consistent null space projections, for example an adaptive one with $\boldsymbol{k}(\boldsymbol{q}, \boldsymbol{q}_0, \boldsymbol{\sigma}) \in \mathbb{R}^n$, where $\boldsymbol{\sigma} \in \mathbb{R}^n$ is the stiffness adjuster of a variable stiffness mechanism [PDAS15].

4.3 Comparison of Null Space Projectors

This section will provide simulations (Sect. 4.3.1) and experiments (Sect. 4.3.2) to demonstrate the properties of the null space projectors in action. In the first simulation, comparisons between successive and augmented null space projections as well as between statically consistent and dynamically consistent redundancy resolutions are made. The second simulation shows the properties of the stiffness-consistent null space projector in comparison with common statically consistent and dynamically consistent redundancy resolutions. In the experimental part the null space projectors are applied to one of the lightweight arms of Rollin' Justin. A discussion of all approaches is provided in Sect. 4.3.3.

4.3.1 Simulations

Simulation #1 demonstrates the theoretical properties of the presented null space projections on a planar $n = 4$ DOF manipulator as depicted in Fig. 4.1. The task hierarchy is designed according to the priority levels specified in Table 4.1.

Since $\sum_{i=1}^{4} m_i = 7 > n$ and the tasks partially conflict with each other, not all of them can be accomplished to the full extent. The controller gains are specified in Table 4.2. In the following, the regulation case and the transient responses are investigated. Figure 4.2 depicts the step responses for five different implementations. Additionally, the solution without any null space projection is plotted. That means that the control torques are directly applied without being processed by any projectors at all, i.e. they are simply added such that all tasks compete with each other without a proper hierarchy.

All augmented methods reach zero steady-state errors on the first three levels because these tasks can be achieved simultaneously. The condition of feasibility can be mathematically written as the existence of a set

$$\mathcal{A} = \left\{(\boldsymbol{q}, \dot{\boldsymbol{q}}) \,|\, \dot{\boldsymbol{q}} = \boldsymbol{0}, \boldsymbol{x}_{i,\text{des}} = \boldsymbol{f}_i(\boldsymbol{q}) \text{ for } i = 1, 2, 3\right\}, \tag{4.52}$$

where $\boldsymbol{x}_{i,\text{des}}$ is the corresponding desired task value of the task vector $\boldsymbol{x}_i$ defined in (4.1). The fourth task, however, cannot be accomplished completely because no set exists which additionally fulfills $\boldsymbol{x}_{4,\text{des}} = \boldsymbol{f}_4(\boldsymbol{q})$. But it is executed as well as possible in a locally optimal way according to the remaining null space. It is noticeable that the steady state is reached considerably later in case of the static null space projections.

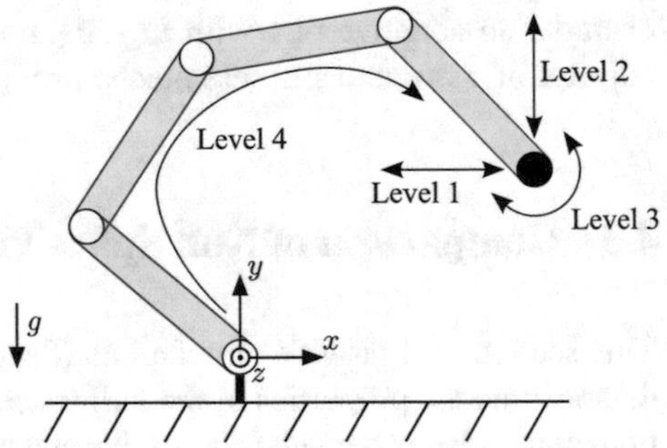

Fig. 4.1 Simulation #1 model of the planar, four-DOF system. The links are connected via revolute joints. Each link is modeled by a point mass of 1 kg that is placed in the center of a bar with length 0.5 m. The dynamics are simulated using $g = 9.81\,\text{m/s}^2$

Table 4.1 Hierarchy for the comparative simulation #1 of different null space projectors

Priority	m_i	Description
$i = 1$	1	Cartesian impedance for the TCP translation in x-direction
$i = 2$	1	Cartesian impedance for the TCP translation in y-direction
$i = 3$	1	Cartesian impedance for the TCP orientation about the z-axis
$i = 4$	4	Full joint impedance

Table 4.2 Controller gains for the simulations and experiments; (* plus additional integrator due to steady-state error)

Gain	Sim. #1	Sim. #2	Experiment #1
$\boldsymbol{K}_1$	$800\ \frac{\text{N}}{\text{m}}$	**0**	$\text{diag}(1200, 1200, 1200)\ \frac{\text{N}}{\text{m}}$
$\boldsymbol{D}_1$	$60\ \frac{\text{Ns}}{\text{m}}$	**0**	damping ratios set to 0.9
$\boldsymbol{K}_2$	$800\ \frac{\text{N}}{\text{m}}$	$200\ \frac{\text{Nm}}{\text{rad}}$, *	$\text{diag}(60, 60, 60)\ \frac{\text{Nm}}{\text{rad}}$
$\boldsymbol{D}_2$	$60\ \frac{\text{Ns}}{\text{m}}$	$10\ \frac{\text{Nms}}{\text{rad}}$	damping ratios set to 0.9
$\boldsymbol{K}_3$	$150\ \frac{\text{Nm}}{\text{rad}}$	–	$\text{diag}(20, \ldots, 20)\ \frac{\text{Nm}}{\text{rad}}$
$\boldsymbol{D}_3$	$4\ \frac{\text{Nms}}{\text{rad}}$	–	$\text{diag}(3, \ldots, 3)\ \frac{\text{Nms}}{\text{rad}}$
$\boldsymbol{K}_4$	$100\ \frac{\text{Nm}}{\text{rad}}$	–	–
$\boldsymbol{D}_4$	$4\ \frac{\text{Nms}}{\text{rad}}$	–	–

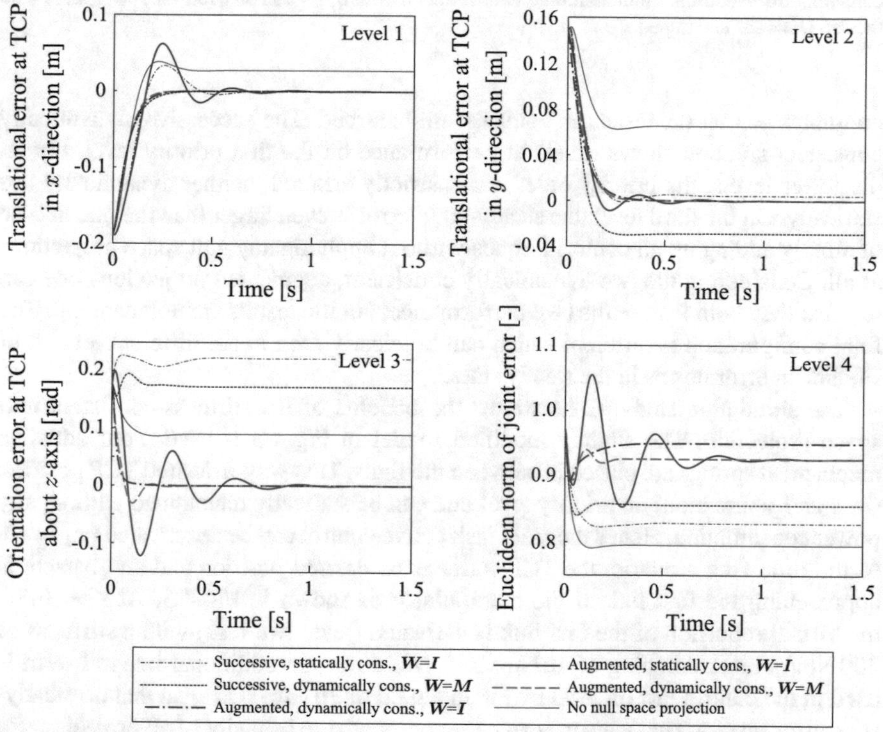

Fig. 4.2 Simulation #1 of different torque control null space projections on a four-DOF manipulator with four individual hierarchy levels

Due to the dynamic coupling of the tasks, disturbing accelerations are generated across the priority levels and slow down the transient behavior. Dynamically consistent null space projectors fulfilling Definition 4.2 implicitly annihilate these inertia

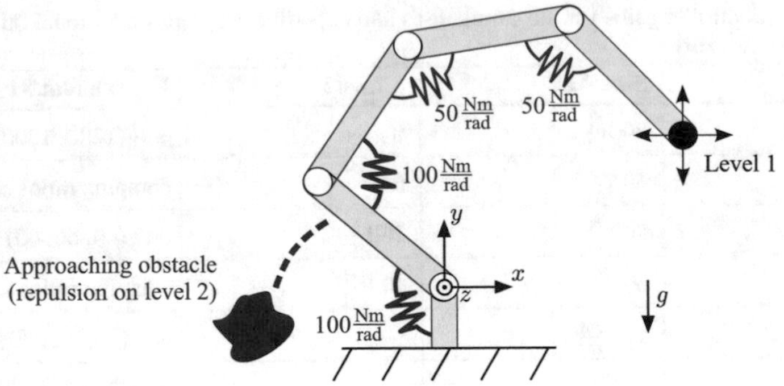

Fig. 4.3 Simulation #2 model of the planar, four-DOF system. The links are connected via revolute joints. Each link is modeled by a point mass of 1 kg that is placed in the *middle* of a bar with length 0.5 m. The dynamics are simulated using $g = 9.81\,\mathrm{m/s^2}$. Four mechanical springs are placed in between the links. These allow to maintain a TCP position without active control and power consumption. Additional joint damping is introduced with $d_i = 15\,\mathrm{Nms/rad}$ for $i = 1, 2, 3, 4$ so that no DOF are undamped

couplings so that the tasks can converge undisturbed. The successive, dynamically consistent solution shows excellent performance on the first priority level, but on the lower levels, the priority order is not strictly ensured, neither dynamically nor statically. On the third level, the steady-state error is even larger than the one in case of simply adding up all control torques without applying any null space projections at all. Considering the two dynamically consistent, augmented projections one can say that they both feature the best performance, but the results are not identical. The final configuration is different which can be clearly seen in the different level four Euclidean error norms in the steady state.

The simulation study #2 illustrates the benefits of the stiffness-consistent null space projection. The slightly modified model in Fig. 4.3 is used. Four adaptive mechanical springs are placed in between the links. That way, a desired TCP position (in x and y direction) on priority level one can be statically maintained without any power consumption. Hence the main task active control can be deactivated ($\boldsymbol{\tau}_1 = \mathbf{0}$). In the following scenario, the TCP starts at its desired position and an obstacle is approaching the first link of the manipulator as shown in Fig. 4.3. At $t = 0.5\,\mathrm{s}$, the virtual repulsion of the first link is activated (level two task) with a stiffness of 200 Nm/rad and damping of 10 Nms/rad. Moreover, an additional integral term is used in the control law on level two with a gain of 10 Nm/(rad s) so that no steady-state error results. That way one can better compare the behavior of all projectors for the same null space control quality, i.e. no steady-state errors remain on level two after the transient. In the upper two diagrams in Fig. 4.4, the Cartesian errors at the TCP are depicted. As shown in Sect. 4.2.3, a stiffness-consistent null space projection minimizes the main task level error in a static sense. The plots reflect these theoretical results. Using $\boldsymbol{W} = \boldsymbol{K}(\boldsymbol{q}_0)$, a small noteworthy error can be observed during the transient, which was expected since the null space projector is of static nature only.

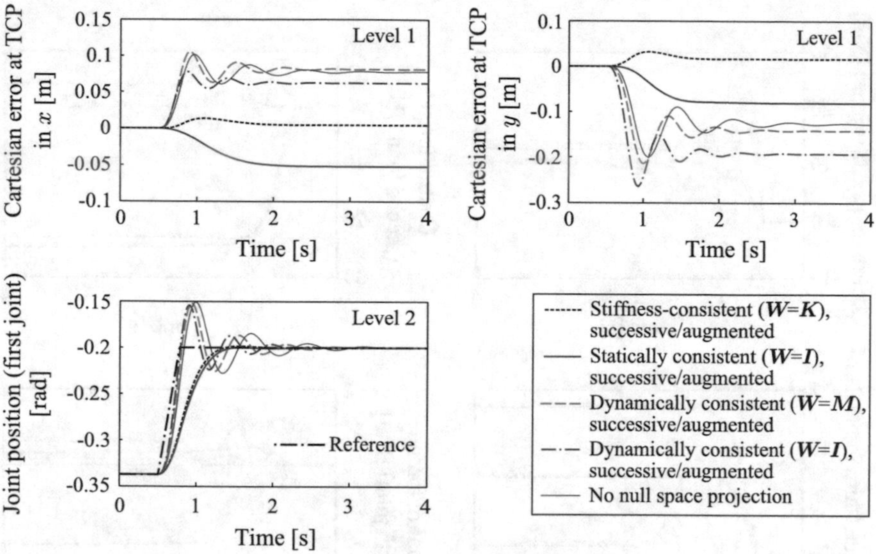

Fig. 4.4 Simulation #2: benefits of a stiffness-consistent null space projection

Although featuring the best performance by far, the stiffness-consistent approach also shows a small steady-state error. This is due to the change in the gravity torques because of the large motion in the null space, cf. (4.50). However, this small error could be easily treated by slight active control w. r. t. the Cartesian space of the TCP. On a real robot, one would certainly activate such an additional control on the first priority level to compensate for inevitable disturbances and model uncertainties but still let the springs do most of the work. It is striking that the dynamically consistent projectors perform very poorly during the transient although they use knowledge of the dynamic capabilities of the system by applying the inertia matrix for the null space determination. But the missing knowledge about the additional springs even leads to worse results than the pure statically consistent projector with $\boldsymbol{W} = \boldsymbol{I}$. Summarized, the comparison with the other null space projectors clearly reveals the advantages of the new concept of stiffness-consistent projectors for this subclass of robots. Note that due to the use of only two priority levels, there is no difference between successive and augmented null space projections. In the bottom chart in Fig. 4.4, the joint position of the first joint is depicted as well as the reference value for the respective secondary task collision avoidance. To get an insight into the applied joint torques, Fig. 4.5 illustrates the control inputs in all joints. One can easily see that the information contained in $\boldsymbol{K}(\boldsymbol{q}_0)$ leads to completely different joint torques, and the final steady state is reached considerably faster compared to most of the other approaches. The control inputs in Fig. 4.5 explain why the dynamically consistent solutions lead to oscillations during the transient, cf. Fig. 4.4. The dynamically consistent redundancy resolutions compensate for the error on level two faster. However, the mechanical springs act like disturbances in these cases because the null space projectors do

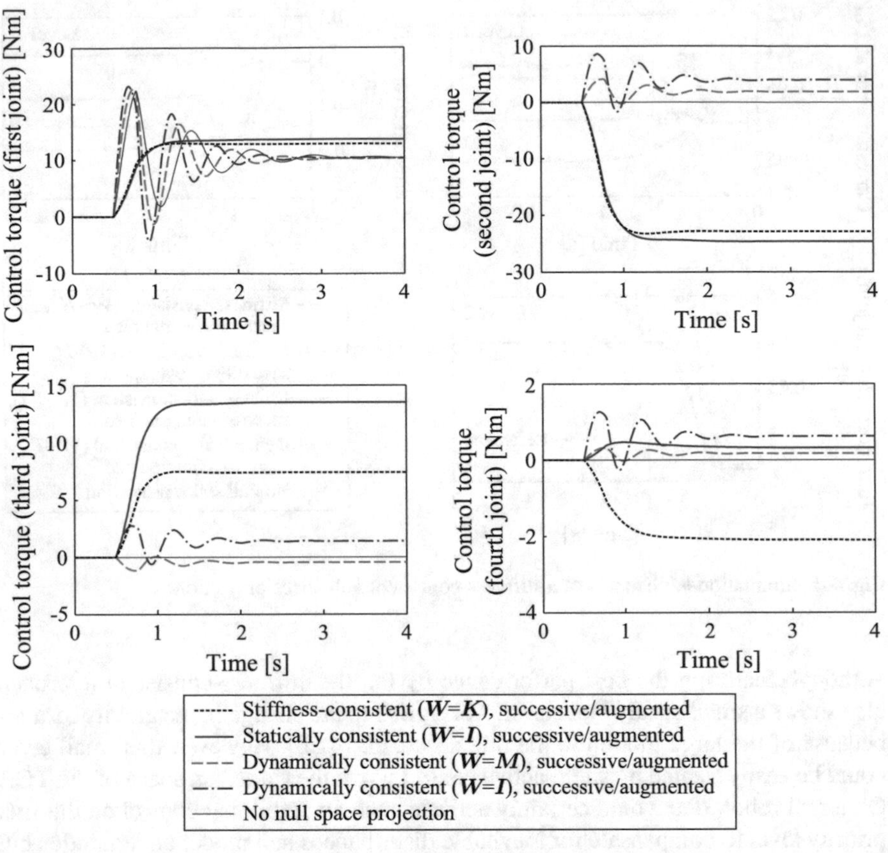

Fig. 4.5 Simulation #2: control torques in the four joints of the simulation model with parallel springs

not take their influence into account. Therefore, the subtask control and the springs compete with each other and lead to a completely different overall stiffness, which alters the transient behavior in turn. The stiffness-consistent null space projector takes the springs into account and avoids such a competition between controller and mechanical springs. Solely the statically consistent null space projector with $\boldsymbol{W} = \boldsymbol{I}$ has a comparably high convergence speed. This is due to the fact that the stiffness matrix in this simulation example is of diagonal shape and thus closer to the identity matrix than the weighting matrices in the other approaches.

4.3.2 Experiments

In the following experiments, the null space projectors are evaluated and compared on the right arm of Rollin' Justin. The task hierarchy is designed as follows:

1. Level ($m_1 = 3$): translational Cartesian impedance at the TCP in x-, y-, z-direction to keep the initial Cartesian position in space (x: forward/backward, y: left/right, z: up/down),
2. Level ($m_2 = 3$): Cartesian impedance for the orientation of the TCP about the three axes with commanded trajectory,
3. Level ($m_3 = 7$): complete joint impedance to maintain the initial joint configuration.

The controller gains are given in Table 4.2. A fast trajectory on the second priority level is applied. Within less than 0.7 s, the TCP orientation is commanded to an intermediate state. After a short rest, it is commanded back to the initial state. The trajectory for the rotation is specified such that its realization requires large motions in the joints of the manipulator. That allows different fundamental aspects to be evaluated in one experiment:

- To what extent is the main task on level one disturbed by control actions on level two and three?
- How good is the task performance on level two due to the restrictions imposed by the task on level one?
- How good is the task performance on level three due to the conflicts with the task on level two?

The performance of the null space projectors can be compared on the basis of Fig. 4.6. The first issue to notice is the clear instability of the augmented, dynamically consistent null space projector with $\boldsymbol{W} = \boldsymbol{I}$ from Sect. 4.2.2.2. At $t \approx 1.2$ s, the emergency stop is used. Although this null space projector has the theoretical advantages shown before, it destabilizes the system. Indeed, that is caused by the procedure

$$\text{torque} \xrightarrow{\boldsymbol{M}(\boldsymbol{q})^{-1}} \text{acceleration} \xrightarrow[\text{null space projection}]{\boldsymbol{I} - \boldsymbol{J}_{i-1}^{\text{aug}}(\boldsymbol{q})^{W+} \boldsymbol{J}_{i-1}^{\text{aug}}(\boldsymbol{q})} \text{acceleration} \xrightarrow{\boldsymbol{M}(\boldsymbol{q})} \text{torque}$$

described in Sect. 4.2.2.2. If $\boldsymbol{M}(\boldsymbol{q})$ has a very small eigenvalue, then $\boldsymbol{M}(\boldsymbol{q})^{-1}$ will have a very large one, i.e. its inverse. If the current torque to be projected has a contribution in the direction of the corresponding eigenvector, the acceleration vector will be "aggressively" scaled. In the second step, the null space projection is performed in the acceleration domain. The projector does not use any information about $\boldsymbol{M}(\boldsymbol{q})$ since $\boldsymbol{W} = \boldsymbol{I}$. In other words, the acceleration vector is projected and the resulting acceleration points into another direction while still suffering from the scaling performed in the first step. In the third step, one goes back to joint torques, but the previous scaling is not reversed. Summarized, one can say that this null space projector "aggressively" scales a torque, depending on the current joint configuration and the eigenvalues of $\boldsymbol{M}(\boldsymbol{q})$, respectively. The infeasibility of the obtained, projected joint torques then destabilizes the system due to actuator limitations, saturation, and the limited torque control bandwidth. This aspect of instability will be picked up and analyzed further in the discussion in Sect. 4.3.3.

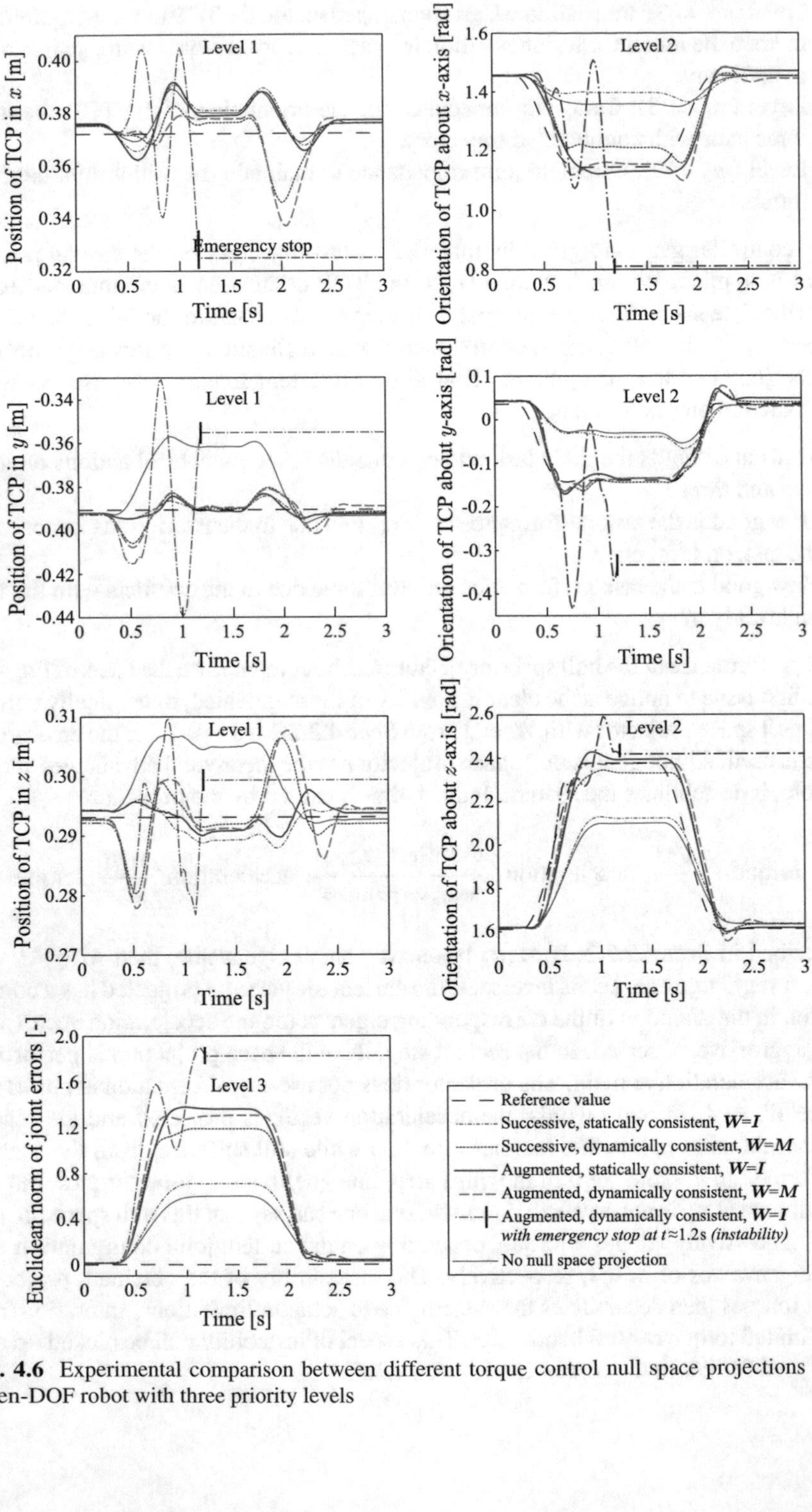

Fig. 4.6 Experimental comparison between different torque control null space projections on a seven-DOF robot with three priority levels

The upper three diagrams on the left side in Fig. 4.6 depict the Cartesian position of the TCP and its reference value. Except for the unstable solution and the summed up control actions ("no null space projection"), the main task is statically achieved. Nevertheless, deviations of several centimeters occur during the transient. Against the expectation of superiority based on the theoretical properties, the projectors using the inertia matrix ($\boldsymbol{W} = \boldsymbol{M}(\boldsymbol{q})$) do not perform better than the projectors without use of it ($\boldsymbol{W} = \boldsymbol{I}$). On the contrary, they generate larger errors in fact. That can be seen in the x- and z-direction at $t \approx 2\,\mathrm{s}$.

As one would expect, the performance on the second level (right column diagrams in Fig. 4.6) is restricted due to the projection onto the null space of the main task. That can be seen in the transient behavior of all three control variables when the desired orientation of the TCP is changed. If the rotational Cartesian impedance was placed on the first priority level instead, the control errors and the overshootings would be smaller for the given parameterization. Furthermore, the plots on the right confirm the theoretical properties of successive null space projections. As in the simulations, they perform worse than the augmented ones due to the non-strict hierarchy they generate. Therefore, the third level task interferes with the second level task and leads to large control errors on level two. That effect can be clearly seen in the rotation about the x-axis and z-axis. But the most remarkable result is that a strict hierarchy (i.e. augmented) does not necessarily require dynamic consistency for high performance during the transient. The comparable performance of the "augmented, statically consistent, $\boldsymbol{W} = \boldsymbol{I}$" solution and the "augmented, dynamically consistent, $\boldsymbol{W} = \boldsymbol{M}(\boldsymbol{q})$" solution in all three directions (right column diagrams in Fig. 4.6) is not in accordance with the theory. Yet it confirms the results from [ASOFH03], where the conclusion was also drawn that the differences between static and dynamic consistency are significantly smaller than expected when the implementation on real hardware is considered. That effect can be traced back to modeling uncertainties (inertia matrix, kinematics, friction) and disturbances, among others. Nakanishi et al. [NCM+08] came to similar conclusions while comparing inertia-weighted redundancy resolutions among each other. The authors stated that the requirement of a highly accurate, estimated inertia matrix is difficult to realize.

On the third level, the successive null space projections perform better than the augmented ones because they do not implement a strict hierarchy. Therefore, the task on the lowest priority level three can be executed using a larger accessible workspace. The stable, augmented solutions ($\boldsymbol{W} = \boldsymbol{I}$, $\boldsymbol{W} = \boldsymbol{M}(\boldsymbol{q})$) show a comparable behavior. They establish a strict hierarchy implying that the task performance on level three will suffer from the limited available workspace. Therefore, it is proper that the largest error norms will be generated with augmented null space projections. Thanks to the different weighting matrices, the steady-state joint configurations are slightly differing as it can be observed at $t = 1.5\,\mathrm{s}$. Nevertheless, since the inertia has no effect in any static configuration, one cannot generalize superiority or inferiority of inertia-based null space projections compared to non-inertia-based solutions in these states.

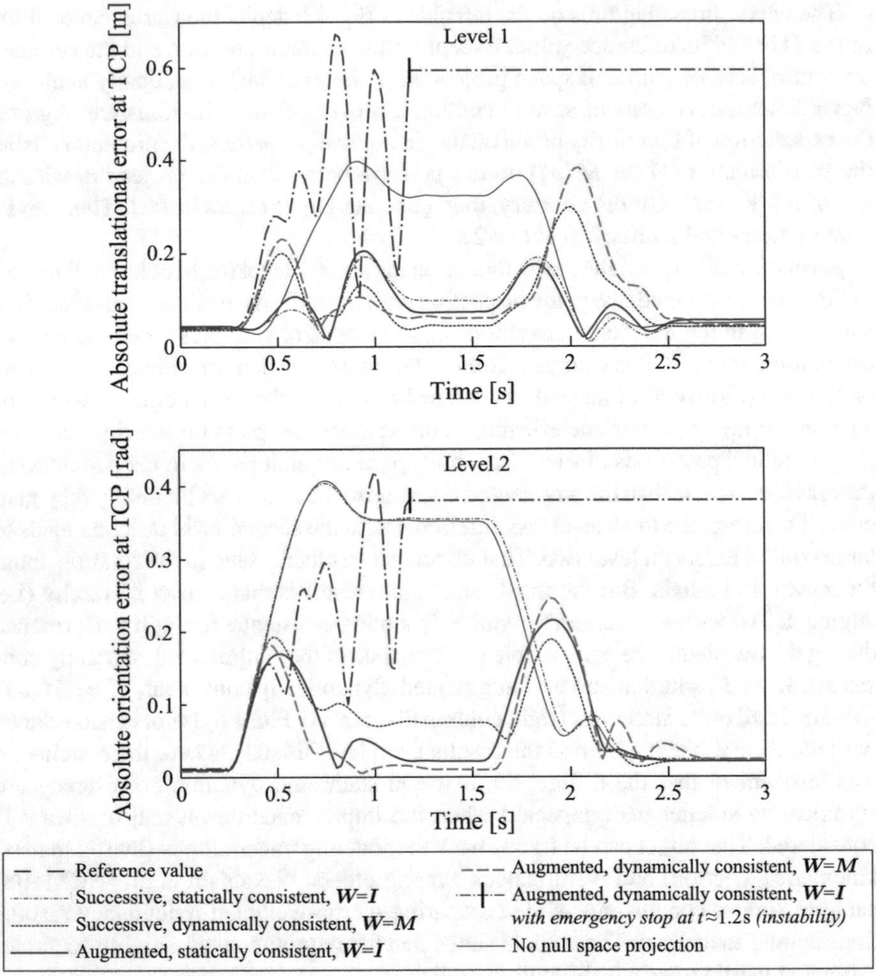

Fig. 4.7 Absolute errors on priority level one and two during the experiments

The total errors in the TCP position and the TCP orientation are plotted in Fig. 4.7. The implemented torque-based tasks realize mechanical impedances. In order to provide the desired physical compliance, the controllers have been implemented following the classical concepts of impedance control via PD-controllers as reported in Sect. 2.2.1. For that reason, small steady-state errors occur. By adding an integral component to the control law, one would erase that error. However, the desired mass-spring-damper behavior, which is beneficial for compliant physical contacts and interaction of the robot with its environment, would be lost then.

Table 4.3 Comparison of different torque control null space projections

	Successive, statically consistent, $W = I$	Successive, dynamically consistent, $W = M(q)$	Augmented, statically consistent, $W = I$	Augmented, dynamically consistent, $W = M(q)$	Augmented, dynamically consistent, $W = I$	No null space projection
Strict hierarchy (static)	Level 1	Level 1	Yes	Yes	Yes	No
Strict hierarchy (dynamic)	No	Level 1	No	Yes	Yes	No
Continuous (no task sing.)	No	No	No	No	No	Yes
Continuous (no algorith. sing.)	Yes	Yes	No	No	No	Yes
Inertia matrix model-free	Yes	No	Yes	No	No	Yes
Idempotent ($NN = N$)	No	No	Yes	Yes	Yes	No
Load independence	No	Level 1	No	Yes	No	No
Stable in experiments	Yes	Yes	Yes	Yes	No	Yes

4.3.3 Discussion

A direct comparison of the null space projectors in terms of their basic properties is given in Table 4.3. The detailed analysis and investigation of the features is presented in the following.

4.3.3.1 Comparison of Successive and Augmented Null Space Projections

The successive null space projection is computationally efficient due to the decoupled calculations of $N_i^{\text{suc}}(q)$. However, a projection onto the null spaces of all higher-priority tasks via (4.6) does not imply strict compliance with the priority order because the tasks are not orthogonal. The matrix $N_i^{\text{suc}}(q)\ \forall\ i > 2$ is not idempotent in general, i.e. the mathematical projection property is not fulfilled due to $N_i^{\text{suc}}(q) \neq N_i^{\text{suc}}(q)N_i^{\text{suc}}(q)$, which is a well-known drawback. The effect on the implementation results can be interpreted easily: A task torque originating from level i is successively multiplied by $i - 1$ matrices obtained by the recursion (4.6). Each multiplication ensures orthogonality to the corresponding higher-level task but it also corrupts all preceding projections at the same time, thus the task hierarchy is not strict in its entirety. Yet, the less complex structure of (4.6) makes it easier to implement dynamic hierarchies such as [DWASH12a], where the priority order can be modified online or tasks get activated and deactivated during operation. The main advantage of the successive projection is that algorithmic singularities are avoided. In the augmented projection such a singularity arises when a rank loss occurs in (4.10), i.e. the tasks on different priority levels completely[5] conflict with each other. Singularities in $\boldsymbol{J}_{i-1}^{\mathrm{aug}}(\boldsymbol{q})$ have to be avoided by smart task definitions or treated by applying singularity-robust techniques such as damped least-squares methods [DW95]. Hence the use of the method complicates the hierarchy design. Nonetheless, the augmented projection enforces orthogonality of all involved tasks, the projection matrix $\boldsymbol{N}_i^{\mathrm{aug}}(\boldsymbol{q})$ always fulfills the idempotence criterion $\boldsymbol{N}_i^{\mathrm{aug}}(\boldsymbol{q}) = \boldsymbol{N}_i^{\mathrm{aug}}(\boldsymbol{q})\boldsymbol{N}_i^{\mathrm{aug}}(\boldsymbol{q})$, thus a strict hierarchy is ensured. In fact, a proof of stability for a generic hierarchy is only known with augmented projections so far [NCM+08, DOAS13].

In successive projections the choice of the weighting matrix cannot solve the problem of a non-strict hierarchy. One has to keep in mind that the type of strictness (successive, augmented) and the kind of consistency (static, dynamic, stiffness) are two independent aspects of the hierarchy design. Thus, a drawback through the choice in the strictness or the consistency cannot be compensated by the choice in the other category. The strictness of the hierarchy determines whether the tasks are properly decoupled or not, and the consistency determines in which way this decoupling is performed, e.g. statically, dynamically, or stiffness-related.

4.3.3.2 Comparison of Static, Dynamic, and Stiffness Consistency

Although dynamically consistent projections have a clear theoretical advantage due to the dynamical decoupling of the priority levels, the final steady state is also achieved with static consistency. Former comparative simulations [CK95] and the ones in Sect. 4.3.1 have shown that the performance of dynamically consistent projections is superior to the statically consistent ones. However, a precise model of the inertia matrix is needed. Experiments on real hardware in Sect. 4.3.2 and [DOAS15] have shown that the differences between the results are significantly smaller than expected. The experimental results given here confirm previous works in the field such as [ASOFH03, NCM+8, PMU+08]. The difference between theoretical superiority and practice can be traced back to modeling uncertainties (inertia matrix, kinematics, friction) and disturbances, for example. In [NCM+08] the authors say that all approaches using the inertia matrix *"significantly degrade, especially in the tasks with fast movements. This implies that these algorithms require highly accurate inertia matrix estimation to be successful"* and they also trace the problems back to inaccuracies in the estimated inertia matrix. In [PMU+08] different redundancy resolution techniques are compared but all of them exploit the inertia matrix either more or less. The authors draw the conclusion that the more influence the inertia matrix has in the control law, the worse the experimental results are. They also report that

[5] In this context, *completely* means that the Jacobian matrix w. r. t. a low-priority task is a linear combination of all higher-level Jacobian matrices. Then this lower-priority task is dropped *completely*, i.e. there is not even a null space left in which it could be partially executed.

the results of simulations are significantly better due to the perfectly known inertia matrix. In [ASOFH03], the first experimental comparison between statically consistent and dynamically consistent null space projections has been made. The results are consistent with the more extensive experiments presented here.

Formal proofs of stability for task hierarchies are intricate [NCM+08] and they are limited to dynamically consistent resolutions so far. In case of two-level hierarchies, see [OKN08, PAW11] for example. A formal proof of stability for a hierarchy with an arbitrary number of priority levels can be found in [DOAS13] or Sect. 5.2, respectively.

In Sect. 4.2.2, two different kinds of dynamically consistent hierarchies have been detailed. The first one is a generalized version of the well-known projector by Khatib [Kha87], which uses the inertia matrix as weighting matrix in the pseudoinversion. Indeed, an infinite number of weighting matrices (4.33) fulfill the same criteria. The second dynamically consistent projector (4.38) refers to a null space projection on acceleration level. The solution was then extended to dynamic consistency by taking the inertia matrix into account in a second step. These two different projectors have very similar theoretical properties as reported in Table 4.3. However, the beneficial property of load independence cannot be concluded for (4.38). Furthermore, severe stability problems have been encountered during the experiments with (4.38). An explanation for the instability has been given in Sect. 4.3.2. The effect is of structural nature and arises from a configuration-dependent scaling from input torque to projected output torque. In configurations where the inertia matrix has one or more small eigenvalues, the null space projection may lead to infeasible joint torques, which exceed the actuator limitations and the torque control bandwidth. Nevertheless, one has to remark that this "aggressive" scaling does not necessarily have to happen, since it depends on the condition of the inertia matrix and the torque to be projected. The simulations in Sect. 4.3.1 have depicted two scenarios in which the closed loop behaved properly when applying the acceleration-based null space projector. The conclusion is that (4.38) is risky to be applied, and since other null space projectors have additional beneficial properties while not suffering from stability issues, there is no convincing reason for the use of (4.38).

It shall also be noted that one can easily obtain a dynamically consistent null space projector without resorting to expensive numerical computations such as singular value decompositions. The only adaptation is to further subdivide all levels from (4.1) and (4.2) such that $m_i = 1 \; \forall i$, which does not pose any problems in general. If a set of equally prioritized tasks is feasible, a strict hierarchy among these subtasks is also feasible. Then the inversion in (4.13) simplifies to the inversion of a scalar. Such a formulation with reduced computational complexity is particularly suitable for real-time applications of dynamic hierarchies where subtasks are activated and deactivated online and the priority order is modified during operation, e.g. by utilizing physically interpretable measures as done in [DWASH12a] and Sect. 4.4.

Stiffness consistency can be interpreted as a subclass of static consistency with particular properties for specific scenarios. Section 4.3.1 demonstrated the advantages of this new null space projector in simulation. In case of mechanical springs placed in parallel to the joints (PEA), a main task can be statically achieved by these passive elements without any power consumption or active control. By applying the stiffness-consistent null space projector, the main task execution through the springs remains undisturbed while a secondary task is executed in its null space. In such a scenario, the stiffness-consistent redundancy resolution is superior to other null space projections.

The aspect of consistency in torque-based null space projections is directly related to the weighting matrix in the pseudoinversion. Except for special cases such as PEA with additional mechanical springs, the weighting matrix primarily has an influence on the transient behavior as the comparison between statically and dynamically consistent null space projections in this chapter has clearly demonstrated. In the equilibrium it is irrelevant which weighting matrix has been chosen because all of them fulfill the definition of static consistency (Definition 4.1). Nevertheless, one has to keep in mind that different transients may lead to different local minima to be reached, so the weighting matrix can also affect the steady-state configuration.

4.4 Unilateral Constraints in the Task Hierarchy

In unstructured and unpredictable environments, robots are compulsorily faced with dynamic hierarchies. Imagine a service robot executing tasks in a kitchen. As soon as a human being enters the room, additional tasks and constraints have to be considered: Collisions with the human have to be avoided, for example. The robot has to observe the person to be able to react to commands. And if necessary, task execution has to be interrupted. All of these subtasks are usually located on different levels in the hierarchy, depending on their importance. In classical null space projections, one would either activate/deactivate the tasks and obtain a discontinuous control law, or redundancy would be "wasted". The latter is due to the fact that these discontinuities can be circumvented by permanently "locking" DOF for momentarily deactivated tasks. But this reservation is at the expense of valuable kinematic redundancy. Hence, a way has to be found to properly deal with these transitions by extending the classical approaches.

In [EC09], Ellekilde and Christensen use the so-called dynamical systems approach to scale task contributions online in case of such a (de)activation. Sugiura et al. blend self-collision avoidance with whole-body motion control to shift the priorities in real time [SGJG07]. Brock et al. propose a dynamic hierarchy, wherein obstacle avoidance is applied in the null space of a primary task [BKV02]. However, it is given a higher priority if that null space reveals not to be sufficient to avoid the collision. A suitable coefficient is calculated online to induce such a transition. In [LMP11], Lee et al. smooth the transition instead of modifying the control law. The proposed framework acts on the kinematic level, joint velocities are the inputs to the robot. Another elaborate, kinematic approach has been proposed by Mansard et al. [MRC09]. They introduce a new inversion operator to ensure continuity and apply it to a visual servoing scenario. An extension for a hierarchy of tasks and unilateral constraints[6] is made in [MKK09]. The extension to the dynamic case is provided in [MK08]. However, it leads to very complex formulations, which are difficult to parameterize. Probably the most common method to deal with discontinuous inverses is to utilize damped least-squares techniques. These singularity-robust inverses (SRI) are widely used in the field of inverse kinematics. A thorough overview is provided by Deo and Walker [DW95]. However, in damped least-squares approaches the proper parameterization of the damping terms is not trivial at all. A simple and intuitive solution, which provides full control over the critical directions and the transition, was still missing so far.

In this section a framework for dynamic hierarchies is presented, which is based on a new and very intuitive formulation of the null space projection [DASH12, DWASH12a]. In contrast to most of the state-of-the-art approaches, the dynamic domain with joint torque interface is considered instead of the kinematic case. The approach manages singular Jacobian matrices, dynamic hierarchies and unilateral constraints. In the latter case, the aforementioned, undesired "locking" can be completely avoided. The method allows to selectively regulate the torque gradient during the transition process, independent of the singular values of the Jacobian matrix. That way, the gap between the abstract mathematical structure of a task hierarchy and the directed influence on real physical values in the robotic system is closed. Discontinuities can be "stretched" and distributed over a well-defined range to comply with any physical constraints. While controlling and fully specifying the exact transition behavior, it is ensured that no deviations from the nominal behavior occur outside the transient phase. The approach is based on a very intuitive interpretation of null space projections and poses no numerical problems when approaching singularities. Only the behavior in the critical directions is altered by the transition shaping while the other directions remain unaffected. Simulations and experimental results show the performance of the redundancy resolution concept. Among others, the self-collision avoidance (Sect. 3.1) has been chosen as an example for unilateral constraints, see Fig. 4.8. The approach is equally applicable for a velocity interface in terms of well-directed limiting of joint accelerations.

4.4.1 Basics

Assume a high-priority task with dimension m which is described by a virtual constraint $\boldsymbol{f}(\boldsymbol{q}) = \boldsymbol{0}$. The Jacobian matrix is $\boldsymbol{J}(\boldsymbol{q}) = \partial \boldsymbol{f}(\boldsymbol{q})/\partial \boldsymbol{q} \in \mathbb{R}^{m \times n}$. Initially, $\boldsymbol{J}(\boldsymbol{q})$ is supposed to be non-singular. In the redundant case ($m < n$), a torque from a lower-priority level can be projected onto the null space of $\boldsymbol{J}(\boldsymbol{q})$ with

$$\boldsymbol{N}(\boldsymbol{q}) = \boldsymbol{I} - \boldsymbol{J}(\boldsymbol{q})^T (\boldsymbol{J}(\boldsymbol{q})^+)^T \tag{4.53}$$

[6] A unilateral constraint describes a task that is not permanently active, but it can be activated and deactivated at run time. An example is given in Fig. 4.8.

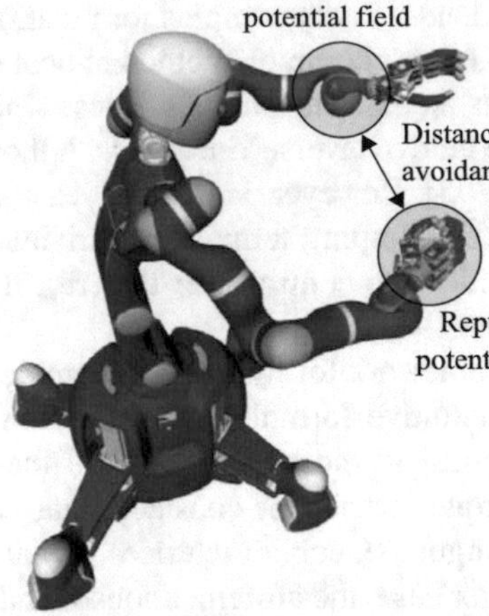

Fig. 4.8 Repulsive potential fields are used to avoid self-collisions. This example shows the contact point pair "*left wrist*"—"*right hand*" on Rollin' Justin. The avoidance gets activated as soon as the potential fields overlap. For that reason, it is named a unilateral constraint

in a statically consistent way (cf. Sect. 4.2.1). Herein, $\boldsymbol{J}(\boldsymbol{q})^{+}$ denotes the Moore-Penrose pseudoinverse of $\boldsymbol{J}(\boldsymbol{q})$, cf. (4.22). From a numerical perspective, the inversion is mostly done by applying a singular value decomposition (SVD) to the Jacobian matrix:

$$\boldsymbol{J}(\boldsymbol{q}) = \boldsymbol{U}(\boldsymbol{q})\boldsymbol{S}(\boldsymbol{q})\boldsymbol{V}(\boldsymbol{q})^{T}. \tag{4.54}$$

The matrices $\boldsymbol{U}(\boldsymbol{q}) \in \mathbb{R}^{m\times m}$ and $\boldsymbol{V}(\boldsymbol{q}) \in \mathbb{R}^{n\times n}$ are orthonormal. The rectangular diagonal matrix $\boldsymbol{S} \in \mathbb{R}^{m\times n}$ contains the singular values $\sigma_1(\boldsymbol{q})$ to $\sigma_m(\boldsymbol{q})$. The pseudoinverse $\boldsymbol{J}(\boldsymbol{q})^{+}$ can be expressed in SVD components:

$$\boldsymbol{J}(\boldsymbol{q})^{+} = \boldsymbol{V}(\boldsymbol{q})\boldsymbol{S}(\boldsymbol{q})^{+}\boldsymbol{U}(\boldsymbol{q})^{T}. \tag{4.55}$$

The inversion of $\boldsymbol{S}(\boldsymbol{q})$ in (4.55) is commonly realized by inverting the diagonal elements and cancelling the singular values below a specified threshold $\varepsilon \in \mathbb{R}^{+}$. At this point, the occurrence of discontinuities becomes evident: If the rank of the Jacobian matrix changes, the threshold ε of one or more singular values will be crossed. That effect propagates back to (4.53) and leads to discontinuities in the control law.

Cancelling the singular values smaller than ε is an arbitrary choice to deal with the singularity while inverting the diagonal elements of $\boldsymbol{S}(\boldsymbol{q})$. Another way to handle that problem is to set a lower bound for the singular values before inverting them. An established method in inverse kinematics is to utilize damped least-squares techniques [DW95] such as

$$\boldsymbol{J}(\boldsymbol{q})^{\dagger} = \boldsymbol{J}(\boldsymbol{q})^{T}(\boldsymbol{J}(\boldsymbol{q})\boldsymbol{J}(\boldsymbol{q})^{T} + \lambda\boldsymbol{I})^{-1}. \tag{4.56}$$

The damped inversion operator is denoted by $\{\}^{\dagger}$. The damping parameter $\lambda \in \mathbb{R}^{+}$ is introduced to smooth the transition by avoiding the division by zero:

$$J(q)^{\dagger} = V(q)\left[S(q)^T\left(S(q)S(q)^T + \lambda I\right)^{-1}\right]U(q)^T \quad (4.57)$$

$$= V(q)\begin{pmatrix} \frac{\sigma_1(q)}{\sigma_1(q)^2+\lambda} & 0 & \cdots & 0 \\ 0 & \frac{\sigma_2(q)}{\sigma_2(q)^2+\lambda} & & 0 \\ \vdots & & \ddots & \frac{\sigma_m(q)}{\sigma_m(q)^2+\lambda} \\ \mathbf{0} & \mathbf{0} & \cdots & \mathbf{0} \end{pmatrix} U(q)^T. \quad (4.58)$$

So far, various different approaches concerning damped least-squares methods have been proposed. First solutions [Wam86] suggested a constant λ but they quickly revealed a crucial problem: Accuracy of the inverse away from the singularity while simultaneously ensuring a smooth transition is hardly feasible. Other solutions used variable damping factors, e.g. dependent on the distance to the singularity [NH86] or its time derivative [KK88]. However, several problems remain. Besides the fact that $J(q)^{\dagger}$ is not a correct inverse of $J(q)$ $\forall\lambda \neq 0$, the choice of the damping parameter is not intuitive and the direct consequence on the projected torques is not obvious.

4.4.2 *Ensuring Continuity*

While activating/deactivating a unilateral constraint or when a singularity is reached, the respective Jacobian matrix changes rank and leads to a discontinuous control law if this issue is not handled properly. The following systematic approach allows to smooth that transition and provides the means to close the gap between the abstract mathematical mechanisms of the projector calculation and the intuitive physics of the robotic system.

4.4.2.1 Intuitive Interpretation of the Null Space Projector

Expressing (4.53) in SVD component notations ((4.54) and (4.55)) leads to the formulation

$$N(q) = I - V(q)S(q)^T U(q)^T \left(V(q)S(q)^{+}U(q)^T\right)^T \quad (4.59)$$

$$= I - V(q)\underbrace{S(q)^T S(q)^{+T}}_{A(q)} V(q)^T. \quad (4.60)$$

Herein

$$A(q) = \operatorname{diag}\left(a_1(q), a_2(q), \ldots, a_m(q), \mathbf{0}_{1\times(n-m)}\right) \quad (4.61)$$

denotes the so-called *activation matrix* with its diagonal elements

$$a_i(\boldsymbol{q}) = \begin{cases} 0 & \text{if } \sigma_i(\boldsymbol{q}) < \varepsilon \\ 1 & \text{otherwise} \end{cases} \quad \forall i,\ 1 \le i \le m. \tag{4.62}$$

An inspection of (4.60) reveals that only the right-singular vectors in $\boldsymbol{V}(\boldsymbol{q})$ have influence on the null space projector,[7] whereas the left-singular vectors in $\boldsymbol{U}(\boldsymbol{q})$ and the exact values σ_1 to σ_m in $\boldsymbol{S}(\boldsymbol{q})$ do not have any influence on the result. The latter can be shown when considering $\boldsymbol{A}(\boldsymbol{q}) \in \mathbb{R}^{n \times n}$. That matrix contains diagonal elements with value 1 (active) or 0 (inactive). The ith diagonal element $a_i(\boldsymbol{q})$ refers to the ith column vector in $\boldsymbol{V}(\boldsymbol{q})$ and either activates that direction or locks it.

Note that (4.60) describes a statically consistent null space projector. Using other weighting matrices in the pseudoinversion, e.g. the joint inertia matrix, would lead to more complex terms that could not be handled as easily. As shown in Sect. 4.3, statically consistent redundancy resolutions feature comparable performance as dynamically consistent ones on real robots.

4.4.2.2 Considering a (1 × 1) Constraint

A $n = 1$ DOF system is illustrated in Fig. 4.9. The depicted mass moves horizontally on the chain-dotted line, the location is described by ϑ. At $\vartheta = \vartheta_0$, a repulsive potential field is penetrated whose purpose it is to prevent a collision with the wall. The potential is the highest-priority task, whereas an arbitrary task defines the mass behavior in the null space of the collision avoidance. Here, the Jacobian matrix of the primary task is

$$\boldsymbol{J} = \sigma_1 \boldsymbol{J}_{1\times 1} \tag{4.63}$$

where the direction $\boldsymbol{J}_{1\times 1} = 1$ is invariant and the singular value σ_1 is extracted from $\boldsymbol{J}$. A SVD of (4.63) leads to (4.54) with $\boldsymbol{U} = 1,\ \boldsymbol{S} = \sigma_1,\ \boldsymbol{V} = 1$. Applying (4.60) delivers

$$N = 1 - SS^+ = 1 - A \tag{4.64}$$

with all matrices degenerated to scalars.

4.4.2.3 Transition for the (1 × 1) Constraint

The discontinuity stated in (4.62) raises the question: Which behavior of the null space projector is actually desired? Evidently, a continuous transition between 0 and 1 is the minimum requirement. Moreover, it is beneficial to limit the projected forces or torques as well as their time derivatives. In the given example, that can be

[7] Actually, only the first m column vectors in $\boldsymbol{V}(\boldsymbol{q})$, which span the subspace of $\boldsymbol{J}(\boldsymbol{q})$, are relevant here. Thus, a reduced SVD suffices to compute the required elements of $\boldsymbol{V}(\boldsymbol{q})$.

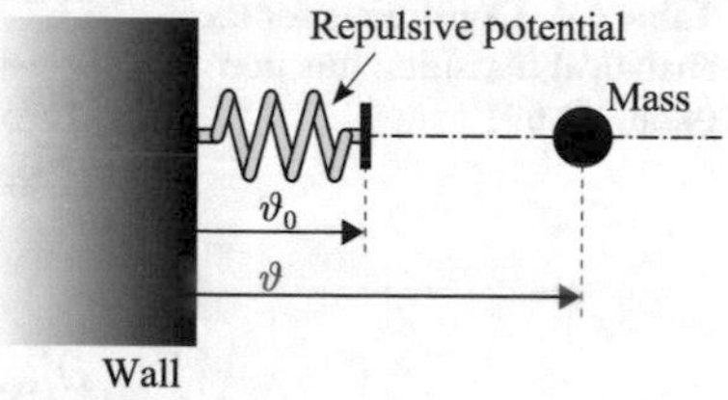

Fig. 4.9 $n = 1$ DOF example of a unilateral constraint: The virtual repulsive potential prevents collisions of the mass with the wall

achieved by shaping the projector such that $N = N_{\text{des}}(\vartheta)$. The variable ϑ does not only determine the position of the mass, but it also indicates whether the unilateral constraint is active or not.

Following (4.64), the activation is defined by

$$a_{1,\text{des}}(\vartheta) = 1 - N_{\text{des}}(\vartheta), \tag{4.65}$$

where $a_{1,\text{des}}$ is the "desired", continuous activation parameter in contrast to the discontinuous one in (4.62). One way to parameterize the transition is to limit $dN_{\text{des}}(\vartheta)/dt$. By projecting a secondary task τ_2 in the null space of the primary task, one obtains the control input

$$\tau_2^{\text{p}} = N_{\text{des}}(\vartheta)\tau_2, \tag{4.66}$$

$$\dot{\tau}_2^{\text{p}} = \frac{\partial N_{\text{des}}(\vartheta)}{\partial \vartheta}\frac{d\vartheta}{dt}\tau_2 + N_{\text{des}}(\vartheta)\dot{\tau}_2. \tag{4.67}$$

In order to get a continuous law with (4.66) and (4.67), $N_{\text{des}}(\vartheta)$ must be of type C^1 at least. In the further analysis, the following assumptions are made:

1. The torque derivative $\dot{\tau}_2$ is neglected. The transition is considerably faster than the variation in the torque τ_2.
2. A maximum or worst case τ_2 can be specified. If that is not possible, an online calculation or measurement is provided.
3. An estimation of the maximum or worst case $\dot{\vartheta}$ is available. If that is not the case, an online calculation or measurement of $\dot{\vartheta}$ is provided.

An analytical expression which allows to specify the transition as described above is the piecewise defined function

$$N_{\text{des}}(\vartheta) = \begin{cases} 0 & \text{if } \vartheta < \vartheta_1 \\ c_3\vartheta^3 + c_2\vartheta^2 + c_1\vartheta + c_0 & \text{if } \vartheta_1 \leq \vartheta \leq \vartheta_2 \\ 1 & \text{otherwise} \end{cases} \tag{4.68}$$

with $[\vartheta_1, \vartheta_2]$ defining the interval from full locking to unconstrained null space projection. Limiting $N'_{\max} = \max\,(\partial N_{\text{des}}(\vartheta)/\partial\vartheta)$ allows to "stretch" the torque variation over a well-defined range. More precisely, a maximum torque derivative $\dot{\tau}^{\text{p}}_{2,\max}$ can be specified, e.g. by referring to the bandwidth of the torque control loop:

Table 4.4 Constraints for the analytical transition function (4.68), (4.69)

	ϑ_1	ϑ_2	$(\vartheta_1+\vartheta_2)/2$
$N_{\text{des}}(\vartheta)$	0	1	
$N'_{\text{des}}(\vartheta)$	0	0	$N'_{\max}$

$$N'_{\max} = \dot{\tau}^{\mathrm{p}}_{2,\max} |\dot{\vartheta}\tau_2|^{-1}. \tag{4.69}$$

The requirements imposed on the analytical transition function are summarized in Table 4.4. The over-determined system of equations can be solved by adding the range $\{\vartheta_2 - \vartheta_1\}$ to the set of unknown parameters $\{c_0, c_1, c_2, c_3\}$ in (4.68).

Notice that limiting $N'_{\max}$ for (4.68) is a conservative way to limit $\dot{\tau}^{\mathrm{p}}_2$ because the maximum slope of $N_{\text{des}}(\vartheta)$ is only reached once within the transition interval. In the following simulation it will be shown that a third-order polynomial is only marginally more conservative than the fastest continuous transition (affine function), but it has the advantage of a significantly smoother behavior. Figure 4.10 shows results for $N'_{\max} = 2,\ 5,\ 20$. The left diagram depicts the activator element $a_{1,\text{des}}$. At $\vartheta = \vartheta_1 = 0.2$, the unilateral constraint is fully activated and the DOF is locked for all low-priority tasks. The parameterization of $N'_{\max}$ can be identified in the right diagram when regarding the maximum slope of N_{des}.

A first-order polynomial has the advantage of a constant $\partial N(\vartheta)/\partial\vartheta$ within the transition phase instead of the quadratic ones shown in Fig. 4.10. Thus, the user-defined $N'_{\max}$ is applied within the whole transition. And it can be shown that the interval size reduces to 2/3. However, a lack of smoothness results at the beginning of the interval and at $\vartheta = 0.2$. Notice also that ϑ_1 does not have to be set equal to ϑ_0 from Fig. 4.9. In the implementations presented subsequently, ϑ_1 is defined as the point of full activation of the primary task. That is a design choice in the hierarchy concept. By specifying $N'_{\max}$, the location of ϑ_2 becomes uniquely determined.

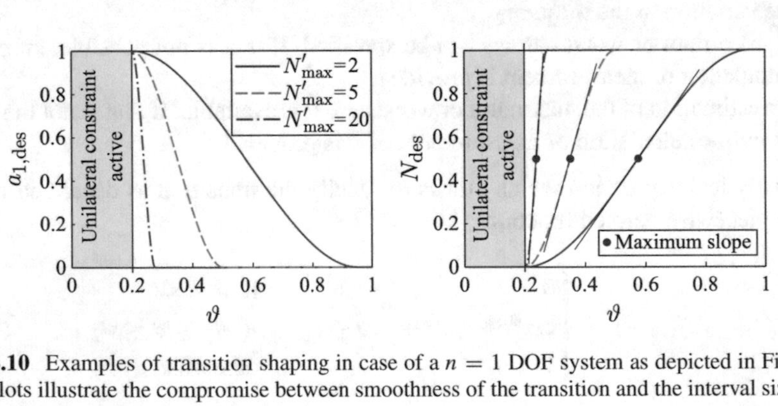

Fig. 4.10 Examples of transition shaping in case of a $n = 1$ DOF system as depicted in Fig. 4.9. The plots illustrate the compromise between smoothness of the transition and the interval size

4.4.2.4 Considering a (1 × *n*) Constraint

The (1×1) case is basically a fade-in and fade-out of the secondary task torque at the activation point of the constraint. Now, an extension to the $(1 \times n)$ case is made with

$$J(q) = \sigma_1(q) J_{1\times n}(q) \in \mathbb{R}^{1\times n}, \quad \|J_{1\times n}(q)\|_2 = 1. \tag{4.70}$$

The original equation (4.60) is modified in a way such that the desired diagonal activation matrix $A_{\text{des}}(\vartheta)$ is used instead of $A(q)$:

$$N(q) = I - V(q) A_{\text{des}}(\vartheta) V(q)^T \tag{4.71}$$
$$= I - J_{1\times n}(q)^T a_{1,\text{des}}(\vartheta) J_{1\times n}(q). \tag{4.72}$$

Only the first diagonal element of A_{des} is important in (4.72). As $J(q)$ is only a row vector, the first column vector in $V(q)$ equals $J_{1\times n}^T$. That turns the method into a computationally efficient technique because only multiplications have to be performed in (4.72) instead of costly decompositions. As reported in Sect. 4.3.3, one can convert any task hierarchy into a stack of one-dimensional tasks so that (4.72) applies.

Variable weights and activators are used in many redundancy resolutions such as [CD95, BKV02, MKK09, LMP11]. However, the purposes and conditions of the activation strongly differ from each other. Chan and Dubey propose a weighted least-norm solution which avoids joint limits [CD95] by means of a configuration-dependent weighting matrix to scale between the different joint contributions. In that approach, the joint limit avoidance is applied on the lowest level where discontinuities never occur. A time-based parameterization for blending and fading out of tasks is proposed by Brock et al. [BKV02]. In contrast to these techniques, the approach proposed here provides direct control over the critical directions via A_{des}. The concept allows to design the transition behavior according to physical limitations of the actuators or the underlying control loops.

4.4.2.5 Transition for the (1 × *n*) and (*m* × *n*) Constraint

One way to handle the complexity in the $(m \times n)$ case is to decompose the lower-level torques τ_2 by projecting them in the critical directions of $V(q)$. The contributions in these critical directions can be used as a basis for the methods from above. Notice that an online decomposition and a feedback into the generation process of $A_{\text{des}}(\vartheta)$ (with several activation states ϑ) closes an additional loop. However, as stated in the first assumption, the transition is supposed to be significantly faster than the variation in the torques from the lower levels. Thus, the effect is expected to be small. An offline consideration is more conservative but does not close a further loop.

The design in the $(1 \times n)$ case is straightforward when applying such an online decomposition because the critical direction is simply $J_{1\times n}$.

4.4.3 Simulations

Simulations have been performed on a planar robot as depicted in Fig. 4.11 (left). It consists of three links and three revolute joints. Viscous joint friction is modeled and the masses are decoupled. A Cartesian impedance is chosen as low-priority task. Its goal is to lead the TCP to the goal configuration (gray dot). The primary task is defined by a singularity avoidance which is designed via a repulsive potential field based on the kinematic manipulability measure (3.55). That singularity avoidance is a unilateral constraint which gets activated if the manipulability measure $m_{\mathrm{kin}}(\boldsymbol{q})$ falls below a specified value $m_0 = 0.23$. The Cartesian reference trajectory of the TCP (Fig. 4.11, left) is designed such that the singularity indicated in Fig. 4.11 (right) is approached. A conflict between the tasks is provoked. Recall that the singular configuration will never be reached by the Cartesian impedance in a steady state. The primary task will outplay the Cartesian impedance because it has a higher priority in the hierarchy. Figure 4.12 depicts the results for the first six seconds of the simulation.

Starting from the initial configuration (Fig. 4.11, left), the TCP moves towards the singular configuration (right). The primary task gets activated at $t = 1.8$ s for the first time. One can observe the transition with $N'_{\mathrm{max}} = 30$ in the upper plot of Fig. 4.12. The second chart shows $N_{\mathrm{des}}(t)$. The primary task (third graph) is active from $\vartheta_2 = 0.23$ on. The bottom diagrams show the Cartesian impedance torques $\boldsymbol{\tau}_2$ and their projections $\boldsymbol{\tau}_2^{\mathrm{p}}$ onto the null space of the primary task.

When the unilateral constraint is fully activated, no Cartesian impedance torque remains in primary task direction in a static sense, see Fig. 4.13. Within the time interval of full activation (shaded rectangle) no torque comes through. Hence, the requirement of an undisturbed priority order is achieved. The control input $\boldsymbol{\tau} = \boldsymbol{\tau}_1 + \boldsymbol{\tau}_2^{\mathrm{p}}$ is depicted in Fig. 4.14 (top). For comparison, a discontinuous null space projection based on a classical matrix inversion is depicted in the bottom chart. Discontinuities can be observed at $t = 1.8$ s, $t = 3.6$ s and $t = 5.4$ s. Applying such commands to a real robotic system would result in unstable behavior as will be shown in one of the following experiments. In this simulation, a steady state in the continuous case is reached after $t = 6$ s asymptotically. In the final configuration, ϑ is a little lower than ϑ_2. No further full transition occurs after $t = 6$ s since the intervention during $1.8\,\mathrm{s} < t < 3.6$ s induced an internal motion that reconfigured the manipulator to comply with the singularity avoidance constraint.

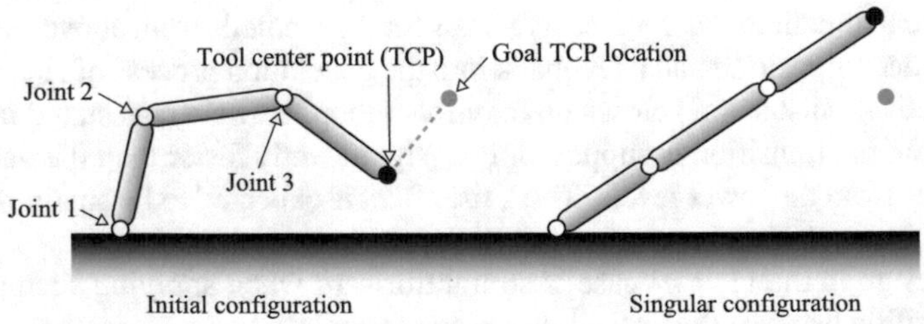

Fig. 4.11 Schematic representation of the planar three-DOF system used for the simulations

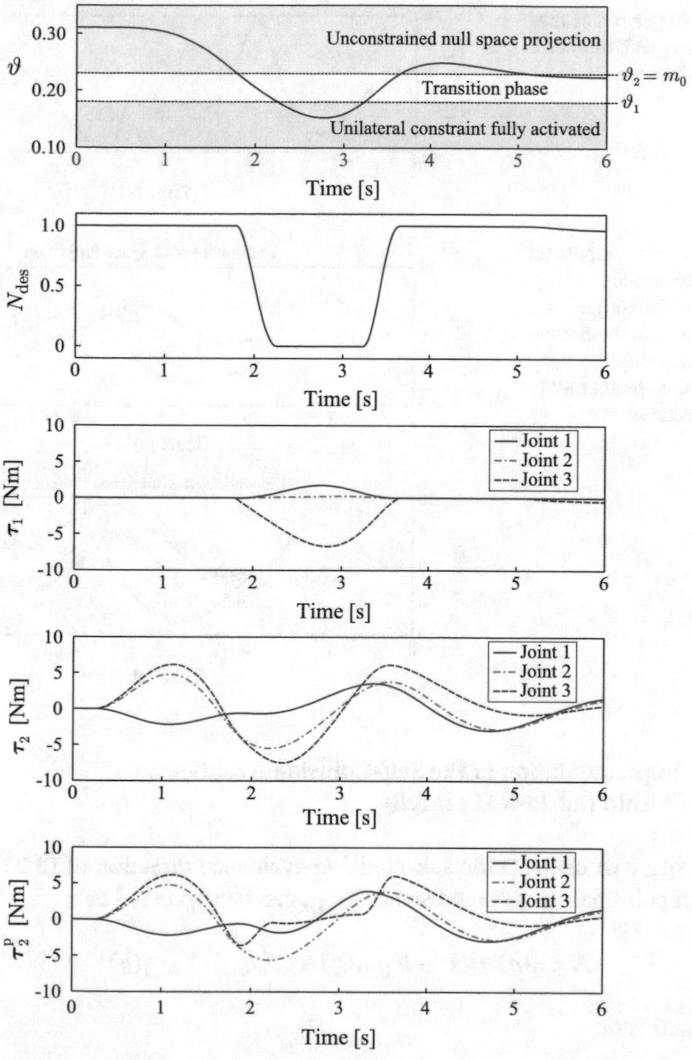

Fig. 4.12 Activation of a unilateral constraint with $N'_{\max} = 30$ in case of a $n = 3$ DOF system simulation. The dynamics are designed with low damping to provoke several penetrations of the transition area. The state of activation of the primary task is given by $\vartheta = m_{\text{kin}}(\boldsymbol{q})$, and the threshold is $\vartheta_2 = m_0 = 0.23$

4.4.4 Experiments

In the first two experiments, the self-collision avoidance (Sect. 3.1) is used to validate the continuous null space projections on a real robot. The third experiment applies a collision avoidance with externals objects to demonstrate the problems of instability.

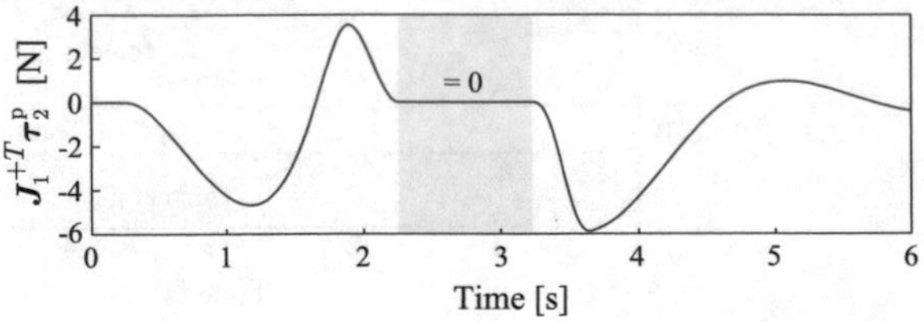

Fig. 4.13 Projection of the lower-priority task torques in the constraint space

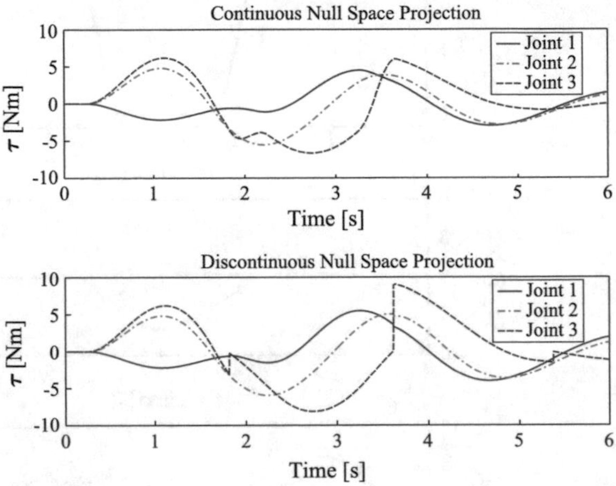

Fig. 4.14 Control inputs for different null space projections. The torque discontinuities in the *bottom* diagram exemplify the problems of common null space projections

4.4.4.1 Implementation of the Self-Collision Avoidance Within the Task Hierarchy

The null space of one specific self-collision avoidance direction of (3.24), defined by contact point pair (i, j) and distance $d_{(i,j)}$, can be expressed as

$$\boldsymbol{N}_{(i,j)}(\boldsymbol{q}) = \boldsymbol{I} - \boldsymbol{V}_{(i,j)}(\boldsymbol{q})\boldsymbol{A}_{\mathrm{des}}(d_{(i,j)})\boldsymbol{V}_{(i,j)}(\boldsymbol{q})^T \tag{4.73}$$

with the activator

$$\boldsymbol{A}_{\mathrm{des}}(d_{(i,j)}) = \begin{pmatrix} a_{1,\mathrm{des}}(d_{(i,j)}) & 0 & \boldsymbol{0} \\ 0 & a_{1,\mathrm{des}}(d_{(i,j)}) & \boldsymbol{0} \\ \boldsymbol{0} & \boldsymbol{0} & \boldsymbol{0} \end{pmatrix} \tag{4.74}$$

of size $(n \times n)$ and

$$\begin{pmatrix} \boldsymbol{J}_i(\boldsymbol{q}) \\ \boldsymbol{J}_j(\boldsymbol{q}) \end{pmatrix} \boldsymbol{V}_{(i,j)}(\boldsymbol{q}) = \begin{pmatrix} \times & \times & \boldsymbol{0} \\ \times & \times & \boldsymbol{0} \end{pmatrix}, \tag{4.75}$$

where $\times$ denotes non-zero elements. Equation (4.75) states that the two Jacobian row vectors are linear combinations of the first two column vectors in $\boldsymbol{V}_{(i,j)}(\boldsymbol{q})$. The projector (4.73) allows many different specifications of the task hierarchy such as

$$\boldsymbol{\tau} = \boldsymbol{\tau}_{\uparrow} + \boldsymbol{N}_{\uparrow}(\boldsymbol{q}) \left(\boldsymbol{\tau}_{\text{sca}} + \left(\prod_{i=1}^{n_{\text{p}}} \boldsymbol{N}_{(i,h(i))}(\boldsymbol{q}) \right) \boldsymbol{\tau}_{\downarrow} \right). \tag{4.76}$$

The subscript $\uparrow$ describes a set of hierarchy levels above the self-collision avoidance, and the subscript $\downarrow$ represents lower-priority levels. Moreover, the projection onto the null space of all higher levels, described by $\boldsymbol{N}_{\uparrow}(\boldsymbol{q})$, may be defined alternatively as for example by the classical approach [SS91].

Instead of (4.76), one can also use recursive algorithms for augmented null space projections as in (4.11)–(4.13). But one has to keep in mind that this requires dealing with algorithmic singularities then.

4.4.4.2 Experiments on Self-Collision Avoidance Within the Task Hierarchy

Only the right arm is active, all other joints are locked, and the priority levels are defined as follows:

Level 1: Self-collision avoidance $\boldsymbol{\tau}_{\text{sca}}$ for the whole upper body of the manipulator and gravity compensation.
Level 2: Six-DOF Cartesian impedance applied to the right TCP.

Experiment #1: The parameterization is given in Table 4.5. The parameters k_{tra} (translational) and k_{rot} (rotational) define the Cartesian stiffnesses which are applied in the three translational and rotational directions. The Cartesian impedance is projected onto the null space of the self-collision avoidance between the left and the right hand. That contact point pair is the most critical one here. Thus, ϑ is chosen to be the distance $d_{(i,j)}$ between these links. The brakes of the left arm are engaged. The initial configuration of the robot is shown in Fig. 4.15a. The snapshots depict the motion of the robot up to the goal location of the right TCP in Fig. 4.15e. The right hand is repelled from the left one during the motion. After reaching the goal location, the right TCP is commanded to move to the initial location again.

Table 4.5 Parameterization for the experiments on self-collision avoidance within the task hierarchy

Experiment	**#1**	**#2**
$F_{\max}$ (N)	30	30
ξ	0.5	0.5
d_0 (m)	0.10	0.10
ϑ_1 (m)	0.05	0.02
ϑ_2 (m)	0.13	0.18
$N'_{\max}$ (1/m)	18.8	9.4
k_{tra} (N/m)	500	500
k_{rot} (Nm/rad)	100	100

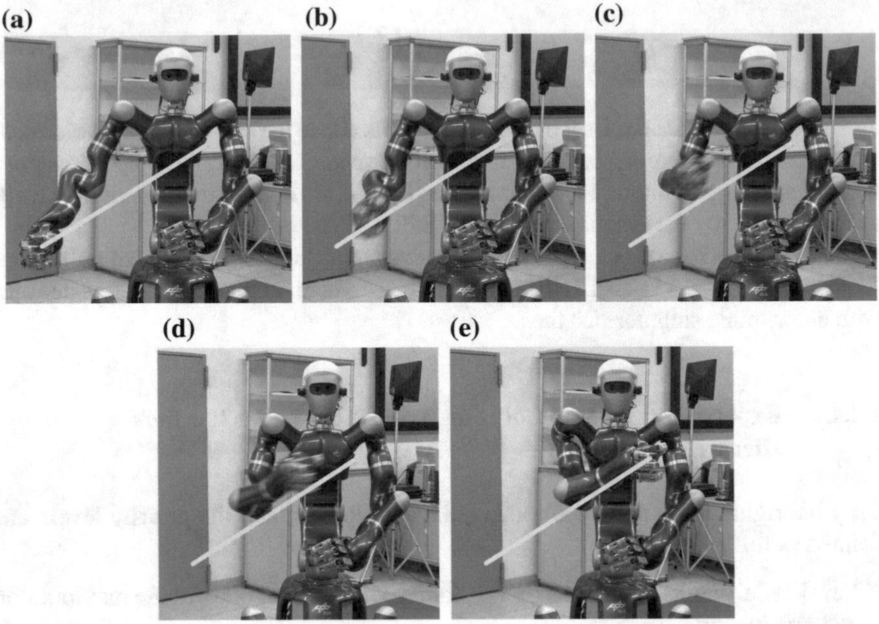

Fig. 4.15 Snapshots during experiment #1: Primary task is the self-collision avoidance between the hands. Secondary task is a six-DOF Cartesian impedance of the right TCP. The trajectory (*solid line*) with a total length of 0.8 m and a maximum translational velocity of 0.6 m/s is realized within 2 s. **a** t = 0.7 s. **b** t = 1.2 s. **c** t = 1.7 s. **d** t = 2.2 s. **e** t = 2.7 s

The self-collision avoidance commands during the motion can be observed in Fig. 4.16 (top). Below, the distance between left and right hand is plotted. Since the avoidance "disturbs" and filters the Cartesian impedance, a deviation between commanded and real TCP location results. The respective translational error is also depicted in this diagram. A higher Cartesian stiffness would reduce the translational error but at the point of complete activation of the high-priority collision avoidance task, a further increase of the stiffness would not have an effect anymore. The direct relation between translational Cartesian error and the penetration of the potential field ($d_{(i,j)}$ hand-hand plot) can be identified easily. In the third chart the transition is depicted. The measured right arm joint torques $\boldsymbol{\tau}_{\text{meas}}$ (bottom) indicate no discontinuities during the transition phase.

The original Cartesian impedance torques of the right arm are provided in Fig. 4.17 (top). The maximum torques in the shoulder and upper arm joints are higher than the ones in the lower arm and wrist joints. That is due to the longer lever arm w. r. t. the right TCP. The second diagram shows the secondary task torques filtered by the null space projector. Notice that the ordinates of the $\boldsymbol{\tau}_2$ plot and the $\boldsymbol{\tau}_2^{\text{p}}$ plot have the same scaling for better comparison. Below, the time derivatives are depicted and feature the desired boundedness. Ignoring the expected noise due to the numerical differentiation, peak values lower than 60 Nm/s can be identified, mainly generated in

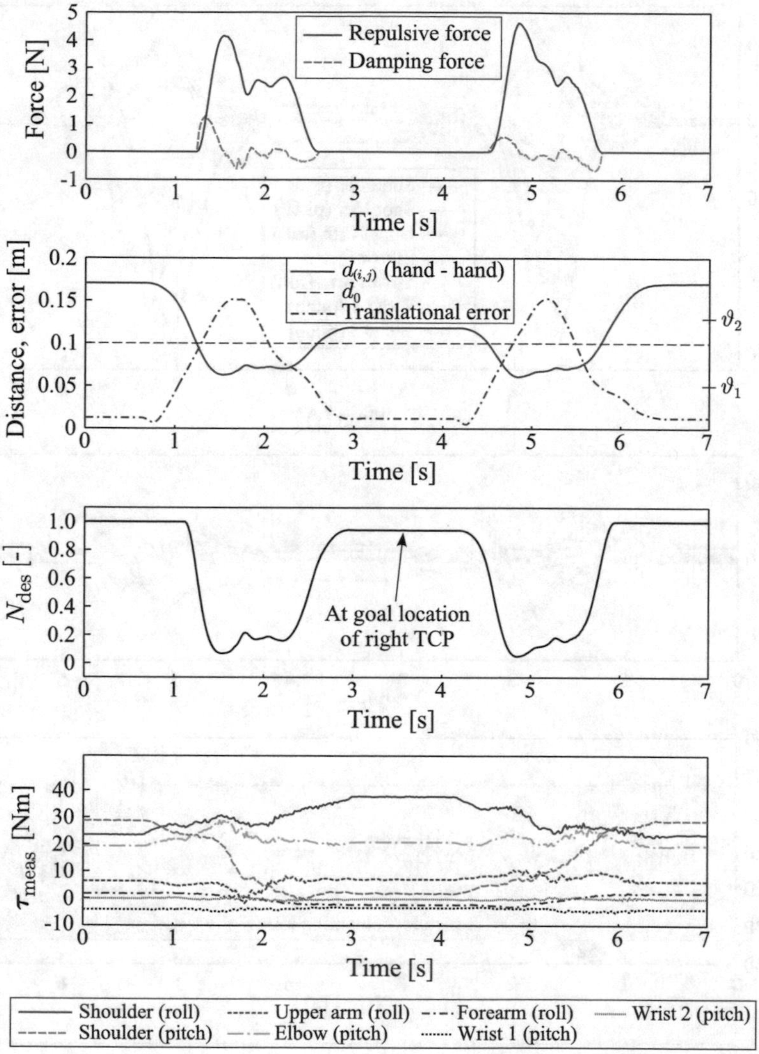

Fig. 4.16 Experiment #1: The secondary task is feasible at the goal location of the right TCP. Data about the self-collision avoidance (top priority) and secondary task execution (Cartesian impedance) is given in the *upper* two charts. The transition behavior is illustrated in the *third* graph. The measured right arm joint torques are provided at the *bottom*

the first arm joints. The curves from elbow to wrist are omitted here and represented by the shaded rectangle instead. The ratio $r_2 = \left\| \boldsymbol{\tau}_2^{\mathrm{p}} \right\| / \left\| \boldsymbol{\tau}_2 \right\|$ at the bottom indicates the instantaneous capability to accomplish the secondary task. While the Cartesian impedance is disturbed in the direction "right hand–left hand", the torques resulting in other directions are unaffected by the null space projection. Notice that the peak at $t = 2.3$ s is only a representation artifact resulting from the normalization because the torque norms are close to zero.

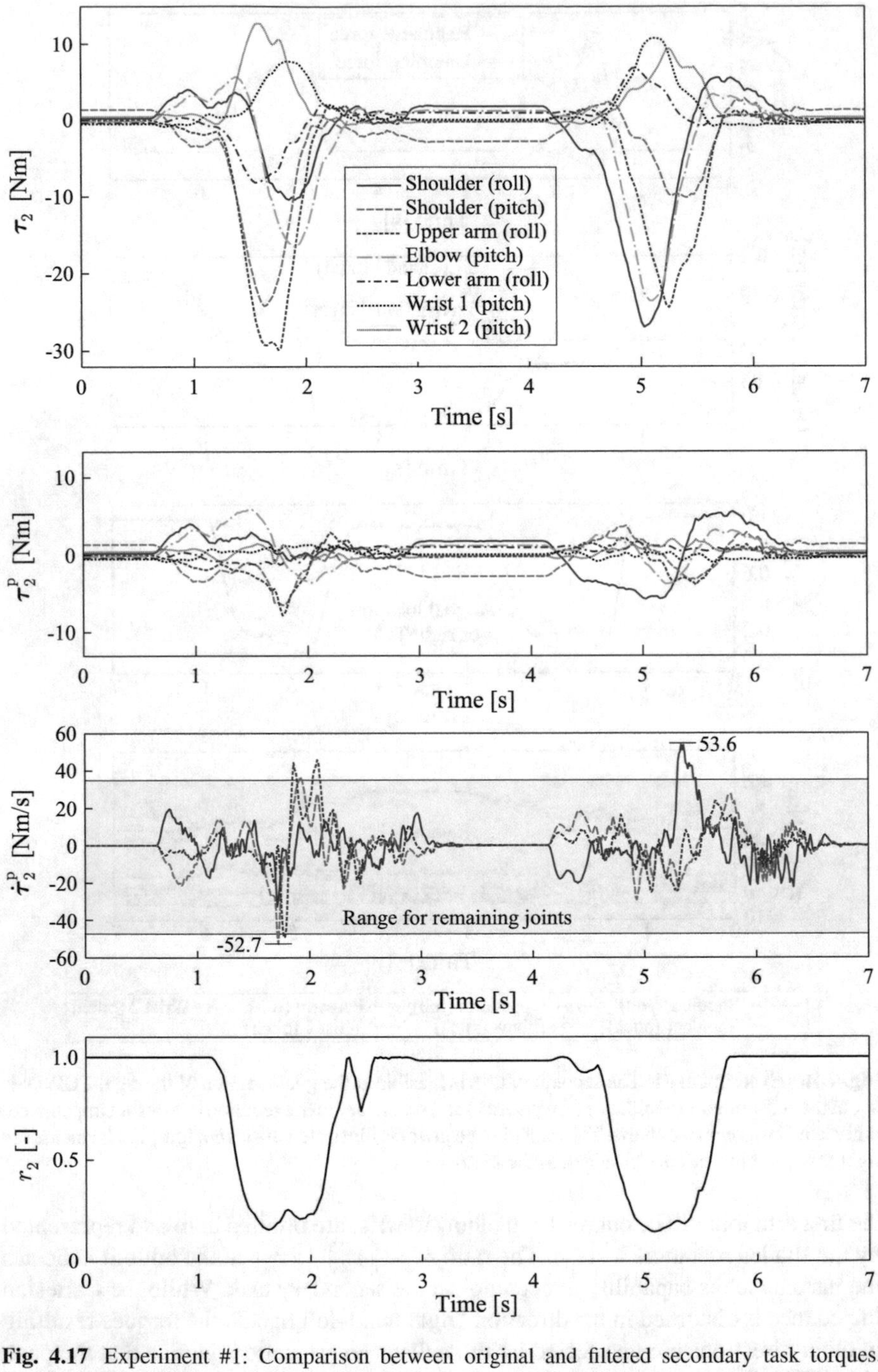

Fig. 4.17 Experiment #1: Comparison between original and filtered secondary task torques. For the differentiation, a low-pass filter with 20 Hz cut-off frequency was applied. The ratio $r_2 = \|\tau_2^p\| / \|\tau_2\|$ indicates the ability to execute the secondary task

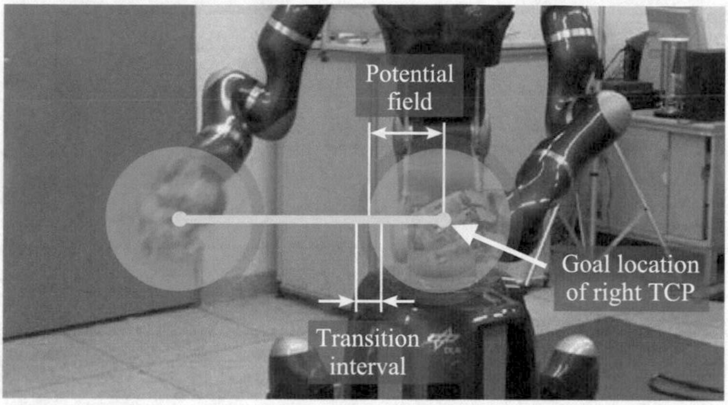

Fig. 4.18 Experiment #2: The *inner circles* represent the self-collision avoidance potential fields of all considered contact point pairs (cpp 1, cpp 2, and cpp 3) qualitatively. In the control, each contact point pair has its own field. The transition interval and the potential field overlap

Experiment #2: A more complex priority order is utilized, based on (4.76). The right TCP is commanded into the left hand as illustrated in Fig. 4.18. After $t = 3$ s, a continuous trajectory, which starts from that intermediate TCP location, leads back to the initial pose. The parameterization is reported in Table 4.5.

The intermediate position is not reached due to the possible hand-hand-collision. Therefore, the Cartesian impedance has to be suspended to comply with the hierarchy. It is projected onto the null space of the most critical self-collision avoidance potentials successively. The priority levels are defined by the contact point pairs of the combinations "right hand–left hand" (cpp 1), followed by "right wrist–left hand" (cpp 2), and "right hand–left wrist" (cpp 3). The top plots in Fig. 4.19 show the activator elements of the three projectors. Although no self-collision avoidance task is completely activated, all of the repulsive potentials partially disturb the secondary task. Notice that they all "work" in different directions, and therefore, they interfere with the impedance multidimensionally. The distances $d_{(i,j)}$ from the collision model are depicted in the second chart. Due to the parameterization of the self-collision avoidance with $d_0 = 0.1$ m, repulsive forces are only generated by the field "right hand–left hand". In this experiment, the repulsion is designed to start in the middle of the transition interval. Thus, the transition begins without a simultaneous self-collision avoidance intervention. That "overlap" is illustrated in Fig. 4.18. The third diagram in Fig. 4.19 shows the Euclidean norms of the Cartesian impedance, its projection via the three null space projectors, and the measured joint torques of the right arm. Most of the secondary task commands are filtered after $t = 1$ s. As expected, the measured torques have a significant offset compared to $\|\boldsymbol{\tau}_2^{\mathrm{p}}\|$ due to the gravity compensation. The bottom plots in Fig. 4.19 give insight into the motion of the active, right TCP. According to the infeasibility of the impedance task, a significant steady-state error remains in the intermediate configuration between $t = 1.5$ s and $t = 3$ s, which follows from respecting the priority order. The deviation from the

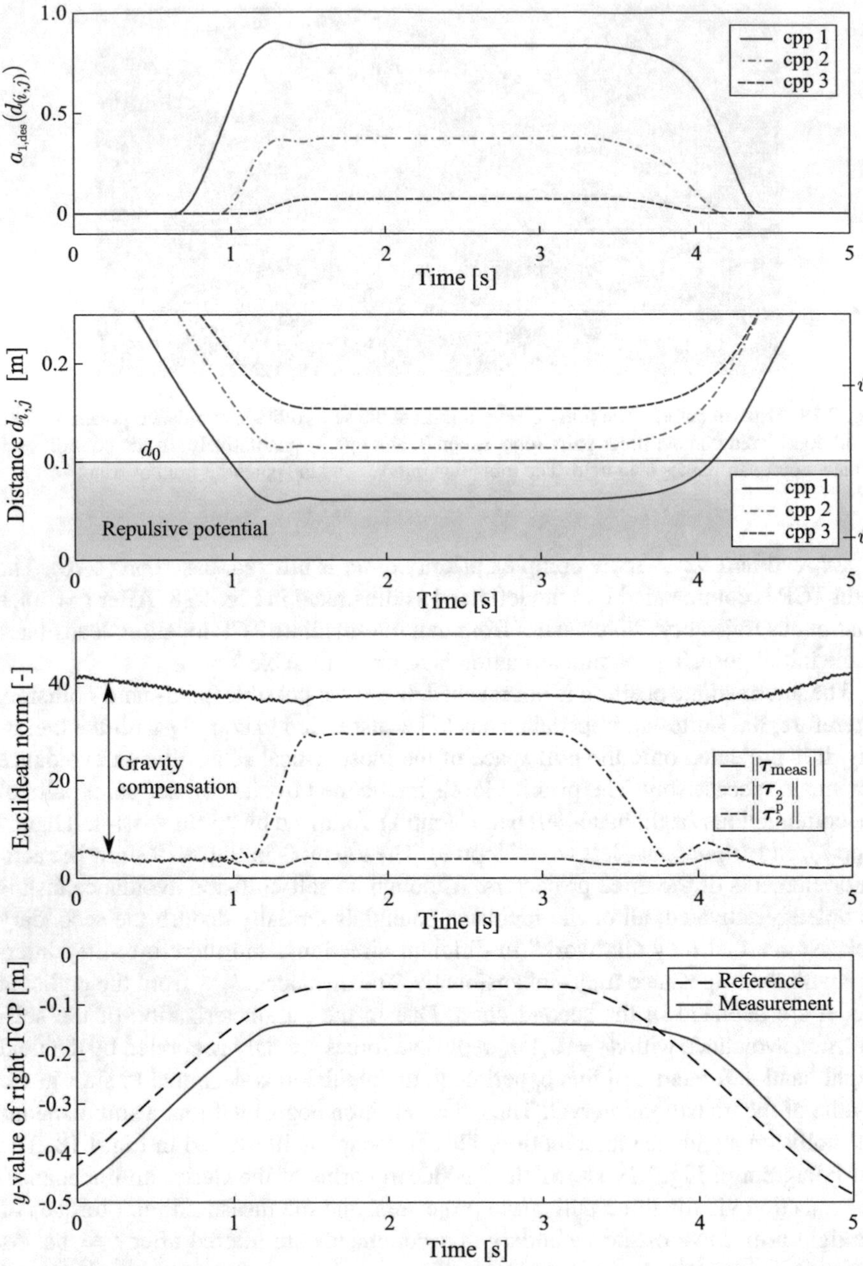

Fig. 4.19 Experiment #2: The secondary task (Cartesian impedance) is not executed completely due to the activation of the primary task (self-collision avoidance). The contact point pairs (cpp) are defined as follows: cpp 1 ("*right hand–left hand*"), cpp 2 ("*right wrist–left hand*"), cpp 3 ("*right hand–left wrist*")

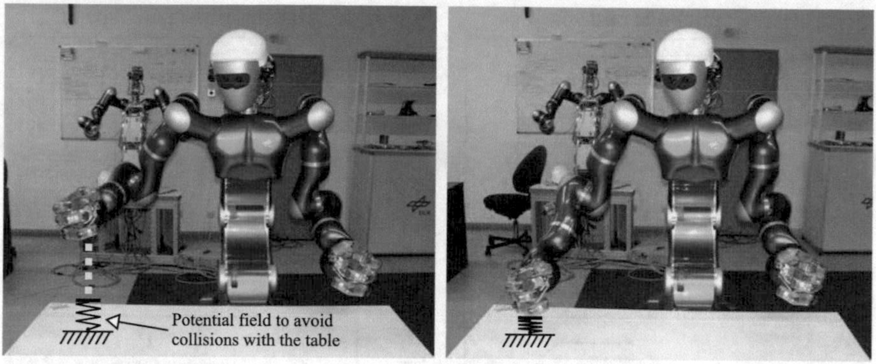

Fig. 4.20 The robot avoids a collision with the table

reference location is also the reason for the slightly increasing $\|\tau_2^{\mathrm{p}}\|$ after $t = 1.3\,\mathrm{s}$ since the Cartesian impedance is not completely deactivated.

Experiment #3: The self-collision avoidance is now replaced by a collision avoidance with the table, see Fig. 4.20. The continuous TCP trajectory for the Cartesian impedance describes a motion of 0.3 m downward along the vertical axis. After 0.25 m, the primary task gets activated which is defined by a unilateral constraint, i.e. a repulsive potential. In the left diagrams of Fig. 4.21, the behavior of the continuous null space is shown. The upper plot depicts $a_{1,\mathrm{des}}$, the bottom plots show the measured torques. Except for some measurement noise, the signals are smooth. The right diagrams illustrate the results for a classical null space projection based on singular value cancellation as described in (4.62). The two tasks compete at the activation border. Particularly within the time interval $7\,\mathrm{s} < t < 8\,\mathrm{s}$, several transitions are triggered, which result in peaks in the torque measurements. Amplitudes up to 50 Nm can be identified. Notice that the plots display the real torques that appear at the joints. The commanded torques attain values of almost 90 Nm but they are not feasible due to the high frequency of the transitions.

4.4.5 Discussion

Compared to scaling and blending techniques such as [EC09], wherein secondary tasks are completely disabled when higher-priority constraints get activated, the *invariance of the remaining directions* is provided here. In other words, only the contributions in the critical directions are influenced. This is due to the property

$$\mathbf{v}_i(\boldsymbol{q})^T \boldsymbol{N} = \begin{cases} \left(1 - a_{i,\mathrm{des}}\right) \mathbf{v}_i(\boldsymbol{q})^T & \text{if } 1 \le i \le m \\ \mathbf{v}_i(\boldsymbol{q})^T & \text{if } m < i \le n \end{cases} \tag{4.77}$$

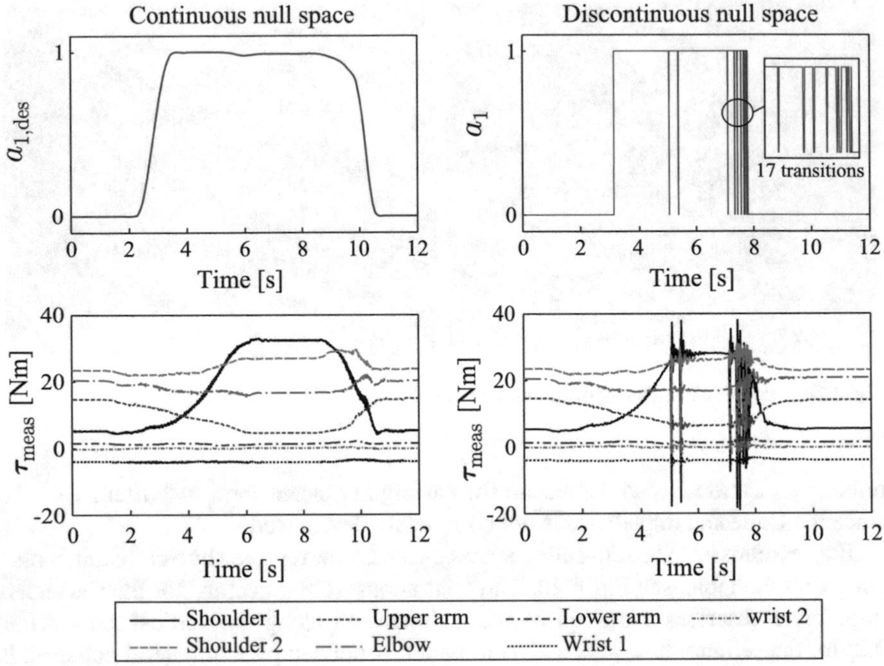

Fig. 4.21 Experiment #3: Different null space projections implemented on the humanoid robot Rollin' Justin. The primary task is a unilateral constraint (collision avoidance), a Cartesian impedance serves as the secondary task. Classical approaches (*right*) destabilize the system

with $\mathbf{v}_i(\boldsymbol{q})$ expressing the ith column vector in $\boldsymbol{V}(\boldsymbol{q})$. The property is a direct consequence of (4.60). The invariance of the remaining $n - m$ uncritical directions is shown in the bottom line of (4.77).

A significant advantage of the concept is the intuitive parameterization and interpretation of the transition. The specification of the transient behavior is the main difference w. r. t. existing approaches that mainly work on a very abstract level [MKK09, LMP11, DSDLR+07]. Adapting to the torque loops allows to exploit the full performance of the hardware. Another major benefit of the approach is its computational efficiency, which is due to the successive null space projection as explained in Sect. 4.1.1. Costly computations such as the SVD can be reduced to a minimum[8] as in (4.73), or they can be completely avoided as in (4.72).

The procedure for unilateral constraints and dynamic task hierarchies is quite similar. In unilateral constraints, tasks are activated or deactivated, respectively. In dynamic task hierarchies, tasks are basically deactivated on one priority level and reactivated on another level. In other words, a dynamic task hierarchy means dealing with unilateral constraints to deliberately alter the order of priority. In both cases, the elements in the activation matrix $\boldsymbol{A}_{\mathrm{des}}(\boldsymbol{q})$ can be chosen as task-dependent, physical

[8] As stated in [BB04], a trimmed-down or reduced SVD is sufficient to compute the projector.

values, e.g. the distance to a collision as exemplified in Fig. 4.9. For the treatment of singularities, one usually does not have such intuitive values. The parameterization of the transition, e.g. via (4.68), can then be specified by employing the actual singular values or the manipulability index (3.55).

4.5 Summary

Chapter 4 addressed the redundancy resolution for robots with multiple simultaneous objectives and a large number of actuated degrees of freedom. It was demonstrated how the reactive methods from Chap. 3 can be arranged in a hierarchy to yield a unified framework for the control of the complete robot.

In Sects. 4.1 and 4.2, the concept of null space projections for priority-based control was surveyed. Different types of implementation were compared from a theoretical and experimental point of view. In the course of this analysis, the classical and widely used dynamically consistent hierarchy was generalized. The solution allows to implement a strict task hierarchy which ensures static and dynamic decoupling of all priority levels. Furthermore, a new kind of null space projector was introduced, namely the so-called stiffness-consistent projection. If the robot is designed with mechanical springs placed in parallel to the joints, this new technique makes it possible to perform secondary tasks without disturbing the main task which is executed by these mechanical springs.

The null space projectors were experimentally compared in Sect. 4.3. It turned out that, once implemented on robots, the theoretically superior dynamically consistent redundancy resolutions yield similar results as the theoretically inferior statically consistent redundancy resolutions. Indeed, statically consistent null space projectors are numerically less expensive and do not require an accurate model of the inertia matrix. Consequently, these findings are of high relevance for robotics.

In Sect. 4.4, the concept of null space projections was enhanced by new features such that the requirements of real-world applications are met, e.g. dynamic environments and conflicting control goals. The proposed solution is able to deal with singular Jacobian matrices, dynamic task hierarchies with online adaptation of the order of priority, and unilateral constraints, which are activated and deactivated on the fly.

The outcome of Chap. 4 is a powerful whole-body control framework to realize various control tasks in a hierarchical order. While the redundancy resolution is able deal with various conditions and requirements, the aspect of stability has not been investigated yet. That point will be addressed in Chap. 5.

Chapter 5
Stability Analysis

This chapter covers the aspect of stability in multi-objective whole-body impedance control. That involves both theoretical stability analyses and the experimental validation of the developed concepts. The chapter is divided into two parts.

In Sect. 5.1, a humanoid robot with torque-controlled upper body and kinematically controlled mobile base is considered. A whole-body impedance is structurally not implementable in a straightforward way due to the kinematically controlled platform. Therefore, an admittance interface is utilized to provide a force-torque input. The objective in Sect. 5.1 is to show stability of the solutions of the dynamic equations of the complete robot [DBP+16, DBOAS14, Bus14]. In Sect. 5.2, the topic of multi-objective control is addressed for robots with force-torque interface. By means of null space projections, a strict task hierarchy can be realized. The proof of asymptotic stability is the first of its kind for a complex task hierarchy in torque control. Section 5.2 is based on [DOAS13, ODAS15].

The results of these two parts are complementary. The outcome of Sect. 5.1 is a stable, wheeled manipulator which can be regarded as a robot with full force-torque interface, where the null space of the main task is not defined yet and can be used for the execution of additional tasks. One can resolve this kinematic redundancy by application of the methods from Sect. 5.2. Asymptotic stability of the desired equilibrium is ensured.

5.1 Whole-Body Impedance with Kinematically Controlled Platform

Mobile humanoid robots with torque-controlled upper body are predestined to be employed in service robotic environments since households (rooms, tools, geometries) are optimized for humans, i.e. two-handed manipulation, human dimensions, and so forth. Unsurprisingly, many complex service tasks have only been executed by wheeled robots so far. Compared to legged humanoid robots, the advantage of

A. Dietrich, *Whole-Body Impedance Control of Wheeled Humanoid Robots*,
Springer Tracts in Advanced Robotics 116, DOI 10.1007/978-3-319-40557-5_5

most wheeled systems is to focus on sophisticated manipulation skills without the necessity of making large efforts for balancing and stabilizing the gait.[1] Based on these considerations, wheeled robots will probably occupy an important role in future service robotics and industrial applications. But a robotics control engineer who wants to implement a whole-body impedance framework as sketched in Sect. 1.4 will face a structural problem: The nonholonomy of the mobile platform requires handling the kinematic rolling constraints for consistent locomotion. These constraints are usually treated by kinematic control of the mobile base on position or velocity level. This is why the control engineer cannot directly access the robot via the force or torque interface needed for impedance control. A solution to that problem is to utilize an admittance interface (cf. Sect. 2.2.2) for the mobile platform.

However, experiments on Rollin' Justin revealed that the whole-body impedance parameters and the platform admittance parameters have to be chosen very conservatively. Instability results otherwise. This conservative parameterization significantly degrades the performance of the method in turn. In Sect. 5.1, these stability issues are analyzed and a control law is presented which leads to a passive closed loop. The convergence of the state to an invariant set is shown with the help of the invariance principle. To prove asymptotic stability in the case of redundancy, priority-based approaches can be employed. Experiments on the humanoid robot Rollin' Justin validate the approach. Section 5.1 is based on [DBP+16, DBOAS14, Bus14] and the outcome can be used for complex manipulation tasks with low-dimensional planning in the task space.

5.1.1 Subsystems

The velocity controller of the mobile base of Rollin' Justin is briefly reviewed, followed by a presentation of the admittance interface. These two subsystems are interconnected such that their combination has a virtual force-torque input. Then the resulting equations of motion are derived including the wheeled platform and the upper body dynamics. Finally, the task space impedance controller is presented and the stability properties are discussed. It is shown that, without modifications, the whole-body impedance results in an unstable system.

5.1.1.1 Mobile Base Velocity Control

The dynamics (2.6) can also be formulated for robots with nonholonomic, wheeled mobile platforms under kinematic rolling constraints. An undercarriage like the platform of Rollin' Justin is called "of type (1, 2)" with a *maneuverability* of dimension $\delta_m = 1$, and a *steerability* of dimension $\delta_s = 2$ [CBDN96, SK08]. Analyzing such

[1]Excluding mobile systems with less than three wheels such as Golem Krang [SOG10] or platforms based on the *Segway* technology.

systems is a standard issue in robotics and will not be covered here. The essence of an investigation of a type (1, 2) platform can be summarized as follows:

1. Although the base cannot change its direction of motion instantaneously, it is able to move freely in the plane by adjusting the wheels (steering and propulsion) appropriately.
2. The mobile platform (of Rollin' Justin) has three DOF for the overall motion, which are: two translations in the plane, and the rotation about the vertical axis. The respective coordinates are denoted $\boldsymbol{r} \in \mathbb{R}^3$. The variable leg lengths of the Rollin' Justin platform (Sect. 3.2) are not considered here.
3. The platform is dynamically feedback linearizable. Giordano et al. [GFASH09] have developed a motion controller which allows to command a desired trajectory $\boldsymbol{r}_{\text{des}}(t)$. An underlying high-gain wheel velocity controller is then employed to realize the necessary wheel behavior (steering and propulsion).

In combination with such an underlying wheel velocity controller, one is able to realize arbitrary desired trajectories in the coordinates $\boldsymbol{r}$, while the kinematic rolling constraints (Pfaffian constraints [SK08]) are complied with automatically. The corresponding control structure is illustrated in Fig. 5.1. Herein, the block *Velocity Controller* includes both the dynamic feedback linearization [GFASH09] and the underlying high-gain wheel velocity controller. The signals $\boldsymbol{w}$ and $\dot{\boldsymbol{w}}$ denote the wheel positions (steering and propulsion) and wheel velocities, respectively. A major feature of the controller is that it compensates for any disturbances thanks to the high gains in the wheel velocity loop. These disturbances are primarily due to dynamic couplings between the upper body of the robot and the mobile base. Moreover, external forces and torques as well as dynamic effects of the platform itself (e.g. inertial forces) are compensated for.

Summarized, the sketched platform velocity control framework leads to the assumption $\dot{\boldsymbol{r}} \approx \dot{\boldsymbol{r}}_{\text{des}}$, while the desired trajectory $\dot{\boldsymbol{r}}_{\text{des}}$ may be arbitrary, provided that it is sufficiently smooth (twice differentiable).

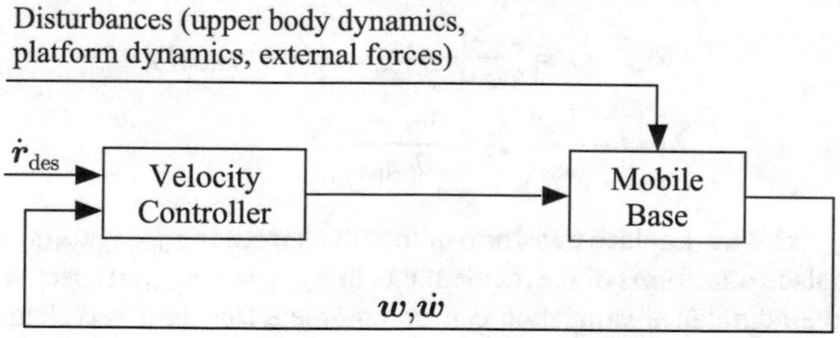

Fig. 5.1 Control loop of the velocity controller for the mobile platform. The control gains are very high in order to compensate for any disturbances

5.1.1.2 Admittance Interface to the Mobile Base

The presented velocity control framework enables to command desired *virtual* base dynamics which the platform is expected to realize. However, since the main goal is to implement a whole-body impedance control law, a force-torque interface is required. Therefore, an admittance is simulated with virtual platform inertia and virtual damping following

$$\boldsymbol{M}_{\text{adm}}\ddot{\boldsymbol{r}}_{\text{des}} + \boldsymbol{D}_{\text{adm}}\dot{\boldsymbol{r}}_{\text{des}} = \boldsymbol{\tau}_{\text{r,vir}} + \boldsymbol{\tau}_{\text{r,ext}}, \tag{5.1}$$

where $\boldsymbol{\tau}_{\text{r,vir}} \in \mathbb{R}^3$ are the virtual forces and torques which can be used as the control input employed by the whole-body impedance to generate the simulated velocity profile $\dot{\boldsymbol{r}}_{\text{des}}$. The external forces and torques $\boldsymbol{\tau}_{\text{r,ext}} \in \mathbb{R}^3$ can only be used in the admittance if the platform is equipped with sensors such that $\boldsymbol{\tau}_{\text{r,ext}}$ can be measured and fed back and/or estimations are available. If no sensors/estimations are available, $\boldsymbol{\tau}_{\text{r,ext}}$ has to be set to zero in (5.1) although external loads may exist physically such that $\boldsymbol{\tau}_{\text{r,ext}} \neq \boldsymbol{0}$ actually holds. Note again that the underlying velocity controller compensates for any disturbances, and the physically exerted $\boldsymbol{\tau}_{\text{r,ext}}$ belongs to this category. The parameters $\boldsymbol{M}_{\text{adm}}$ and $\boldsymbol{D}_{\text{adm}}$ represent the virtual inertia and damping in the admittance, respectively. A reasonable choice for these values is

$$\boldsymbol{M}_{\text{adm}} = \text{diag}(m_{1,\text{adm}}, m_{2,\text{adm}}, m_{3,\text{adm}}), \tag{5.2}$$

$$\boldsymbol{D}_{\text{adm}} = \text{diag}(d_{1,\text{adm}}, d_{2,\text{adm}}, d_{3,\text{adm}}), \tag{5.3}$$

where $m_{1,\text{adm}}, m_{2,\text{adm}} \in \mathbb{R}^+$ describe the virtual platform mass, and $m_{3,\text{adm}} \in \mathbb{R}^+$ is the virtual moment of inertia. Usually one chooses $m_{1,\text{adm}} = m_{2,\text{adm}}$ so that the perceived mass is direction-independent. It seems natural to specify a decoupled damping to determine the positive definite matrix via $d_{i,\text{adm}}$ for $i = 1 \ldots 3$, where $d_{1,\text{adm}} = d_{2,\text{adm}}$. A possible approach to select suitable gains is to consider the Laplace transform of (5.1), which is a first-order low-pass filter:

$$sR_{i,\text{des}}(s) = \frac{K_i}{T_i s + 1}\left(\tau_{\text{r},i,\text{vir}}(s) + \tau_{\text{r},i,\text{ext}}\right), \tag{5.4}$$

$$K_i = \frac{1}{d_{i,\text{adm}}}, \quad T_i = \frac{m_{i,\text{adm}}}{d_{i,\text{adm}}}. \tag{5.5}$$

Here, $R_{i,\text{des}}(s)$ is the Laplace transform of the ith element in $\boldsymbol{r}_{\text{des}}$, $\tau_{\text{r},i,\text{vir}}(s)$ and $\tau_{\text{r},i,\text{ext}}$ are the Laplace transforms of the ith elements in $\boldsymbol{\tau}_{\text{r,vir}}$ and $\boldsymbol{\tau}_{\text{r,ext}}$, respectively. Based on (5.5) the admittance simulation can be parameterized in a very intuitive way. First, one chooses the inertia parameters $m_{i,\text{adm}}$ for $i = 1 \ldots 3$, i.e. the inertia of the platform which is supposed to be perceived. The second intuitive choice is the gain K_i for $i = 1 \ldots 3$, because the whole-body impedance control framework will deliver maximum desired forces and torques (down to the platform), and via K_i the respective maximum admittance velocity can be directly computed. This procedure allows to restrict the base velocities, e.g. for safety purposes. Limiting the forces and torques in the whole-body impedance controller ensures that the error between the actual and the desired TCP position/orientation does not lead to infeasible control inputs due to saturation of the actuators.

Note again that the admittance (5.1) does *not* represent a real physical system but it is only a simulated, desired dynamic behavior the platform is supposed to realize. In other words, if the admittance mass $m_{1,\mathrm{adm}}, m_{2,\mathrm{adm}}$ is set to 10 kg, the platform of Rollin' Justin will behave like it only weighs 10 kg although its real mass amounts to about 150 kg. The active rescaling can be used advantageously to redistribute the apparent inertias so that all body parts of the robot have comparable inertias in order to yield a more natural whole-body behavior.

5.1.1.3 Overall Dynamics

Under the assumption $\dot{\boldsymbol{r}} \approx \dot{\boldsymbol{r}}_{\mathrm{des}}$ from above, the dynamics can be formulated as

$$\begin{pmatrix} \boldsymbol{M}_{\mathrm{adm}} & \boldsymbol{0} \\ \boldsymbol{M}_{\mathrm{qr}} & \boldsymbol{M}_{\mathrm{qq}} \end{pmatrix} \begin{pmatrix} \ddot{\boldsymbol{r}} \\ \ddot{\boldsymbol{q}} \end{pmatrix} + \begin{pmatrix} \boldsymbol{D}_{\mathrm{adm}} & \boldsymbol{0} \\ \boldsymbol{C}_{\mathrm{qr}} & \boldsymbol{C}_{\mathrm{qq}} \end{pmatrix} \begin{pmatrix} \dot{\boldsymbol{r}} \\ \dot{\boldsymbol{q}} \end{pmatrix} + \begin{pmatrix} \boldsymbol{0} \\ \boldsymbol{g}_{\mathrm{q}} \end{pmatrix} = \begin{pmatrix} \boldsymbol{\tau}_{\mathrm{r,vir}} \\ \boldsymbol{\tau}_{\mathrm{q}} \end{pmatrix} + \begin{pmatrix} \boldsymbol{\tau}_{\mathrm{r,ext}} \\ \boldsymbol{\tau}_{\mathrm{q,ext}} \end{pmatrix} . \quad (5.6)$$

The first line represents (5.1), while the second line describes the upper body dynamics with joint configuration $\boldsymbol{q} \in \mathbb{R}^{n_{\mathrm{q}}}$ for n_{q} upper body joint variables. Herein, $\boldsymbol{M}_{\mathrm{qq}}$ is the respective upper body joint inertia matrix, and $\boldsymbol{M}_{\mathrm{qr}}$ is the inertia coupling to the mobile base. Accordingly, $\boldsymbol{C}_{\mathrm{qr}}$ and $\boldsymbol{C}_{\mathrm{qq}}$ denote the corresponding Coriolis/centrifugal terms. The upper body gravity torques are contained in $\boldsymbol{g}_{\mathrm{q}}$, the torques $\boldsymbol{\tau}_{\mathrm{q}}$ are considered as the control inputs to the upper body, and $\boldsymbol{\tau}_{\mathrm{q,ext}}$ are the external forces related to the upper body. Dependencies on the states are omitted in the notations of (5.6) for the sake of simplicity. For later use, the vector

$$\boldsymbol{y} = \begin{pmatrix} \boldsymbol{r} \\ \boldsymbol{q} \end{pmatrix} \quad (5.7)$$

is defined that describes the configuration of the robot. The mobile platform is only represented by its Cartesian position $\boldsymbol{r}$, while the wheel positions and steering angles $\boldsymbol{w}$ do not appear in (5.7), because they are states of the underlying platform control and not relevant in terms of whole-body motions. Therefore, (5.7) is a reduced configuration description, but it defines the actuated DOF which the whole-body impedance controller will access. Before introducing the impedance law, the properties of (5.6) are summarized:

1. A high-gain velocity controller is used to fulfill the rolling constraints and to realize a desired admittance dynamics (5.1) for the mobile base. All disturbances, including the dynamic couplings from the upper body, are assumed to be compensated properly by this controller (cf. Fig. 5.1).
2. The term $\boldsymbol{\tau}_{\mathrm{r,vir}}$ can be used as the control input for the mobile platform.

3. The term $\boldsymbol{\tau}_{\mathrm{q}}$ can be used as the control input for the upper body joints.
4. The properties of the dynamics (5.1) do not comply with the standard rigid body dynamics (2.6) any longer, e.g. the inertia matrix is not symmetric and the property (2.7) cannot be concluded from (5.1) anymore.

5.1.1.4 Impedance Control in the Operational Space

A spatial impedance is designed in the operational space, e.g. the Cartesian space of the TCP. The desired TCP behavior is implemented by applying the impedance to the complete system so that overall compliance is achieved. The spatial error $\tilde{\boldsymbol{x}} \in \mathbb{R}^{n_{\mathrm{x}}}$ in the operational space is given by

$$\tilde{\boldsymbol{x}}(\boldsymbol{y}) = \boldsymbol{x}(\boldsymbol{y}) - \boldsymbol{x}_{\mathrm{des}}, \tag{5.8}$$

where n_{x} is the dimension of the operational space (for full Cartesian impedance: $n_{\mathrm{x}} = 6$), $\boldsymbol{x}(\boldsymbol{y})$ describes the forward kinematics in world coordinates, and $\boldsymbol{x}_{\mathrm{des}}$ is the desired TCP position in world coordinates. The positive definite, virtual potential $V_{\mathrm{imp}}(\tilde{\boldsymbol{x}}(\boldsymbol{y}))$ represents the spatial spring, which may be of the form

$$V_{\mathrm{imp}}(\tilde{\boldsymbol{x}}(\boldsymbol{y})) = \frac{1}{2}\tilde{\boldsymbol{x}}(\boldsymbol{y})^T \boldsymbol{K}\tilde{\boldsymbol{x}}(\boldsymbol{y}), \tag{5.9}$$

for example, with the positive definite stiffness matrix $\boldsymbol{K} \in \mathbb{R}^{n_{\mathrm{x}} \times n_{\mathrm{x}}}$. The gradient of the spring potential and the upper body damping torques are:

$$\boldsymbol{\tau}_{\mathrm{imp}} = -\left(\frac{\partial V_{\mathrm{imp}}(\tilde{\boldsymbol{x}}(\boldsymbol{y}))}{\partial \boldsymbol{y}}\right)^T, \tag{5.10}$$

$$\boldsymbol{\tau}_{\mathrm{damp}} = -\begin{pmatrix} \boldsymbol{0} \\ \boldsymbol{D}_{\mathrm{qq}}\dot{\boldsymbol{q}} \end{pmatrix}. \tag{5.11}$$

That yields the control input

$$\begin{pmatrix} \boldsymbol{\tau}_{\mathrm{r,virt}} \\ \boldsymbol{\tau}_{\mathrm{q}} \end{pmatrix} = \boldsymbol{\tau}_{\mathrm{imp}} + \boldsymbol{\tau}_{\mathrm{damp}}. \tag{5.12}$$

The damping matrix $\boldsymbol{D}_{\mathrm{qq}} \in \mathbb{R}^{n_{\mathrm{q}} \times n_{\mathrm{q}}}$ must be positive definite. Damping is not applied to the platform subsystem in (5.11) because an energy-dissipating term has already been used in the admittance (5.1). The impedance torque $\boldsymbol{\tau}_{\mathrm{imp}}$ consists of n_{q} elements related to the upper body and $n_{\mathrm{r}} = 3$ elements related to the mobile base. The upper body torques can be applied directly and commanded to the torque controllers in the respective joints via $\boldsymbol{\tau}_{\mathrm{q}}$. The elements in $\boldsymbol{\tau}_{\mathrm{imp}}$ related to the platform are commanded as $\boldsymbol{\tau}_{\mathrm{r,vir}}$ in the admittance (5.1).

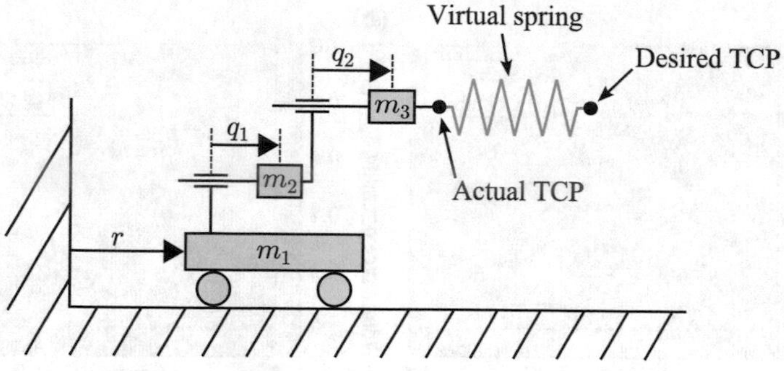

Fig. 5.2 Linear simulation model with three DOF

5.1.1.5 Interaction Between Torque-Controlled Upper Body and Admittance-Controlled Platform at the Example of a Linear Three-DOF System

In the following, a mobile, linear system with three DOF is analyzed to demonstrate the stability problems of (5.6) in combination with the whole-body impedance controller (5.12). The manipulator is sketched in Fig. 5.2. The system parameters and the controller gains are given in Table 5.1. The admittance parameterization varies.

Due to the linearity, the system stability can be analyzed based on the closed-loop poles. For different parameterizations of the virtual platform mass m_{adm} and damping d_{adm}, the maximum real part of all eigenvalues is computed. The areas in Fig. 5.3 show when the system is stable and when it is unstable. The reason behind the differences is the inertia coupling $\boldsymbol{M}_{\text{qr}}$ from (5.6) between upper body and mobile base. When compensating for $\boldsymbol{M}_{\text{qr}}$, stability is ensured for *all* reasonable parameterizations ($m_{\text{adm}} > 0$, $d_{\text{adm}} > 0$). In such a case the dynamic parameters fulfill $\boldsymbol{M} > \boldsymbol{0}$, $\boldsymbol{D} > \boldsymbol{0}$ and all properties of (2.6). In contrast, the resulting inertia matrix is not positive definite anymore in the case of uncompensated $\boldsymbol{M}_{\text{qr}}$. Note that, although no Coriolis/centrifugal couplings between the subsystems "upper body" and "platform" exist due to the use of prismatic joints, the subsystems are always coupled via the spatial spring from the whole-body impedance. That applies to both scenarios, i.e. Fig. 5.3a, b.

Albeit only analyzed for a simple linear system here, similar effects can be observed on Rollin' Justin during experiments. Note that on Rollin' Justin, one has Coriolis and centrifugal couplings additionally, thus $\boldsymbol{C}_{\text{qr}} \neq \boldsymbol{0}$.

Table 5.1 Parameters for the simulation of the mobile robot with three DOF

Parameter	Value	Unit
m_1, m_2, m_3	1	kg
$\boldsymbol{K}$	1	N/m
$\boldsymbol{D}_{\text{qq}}$	diag(1, 1)	kg/s

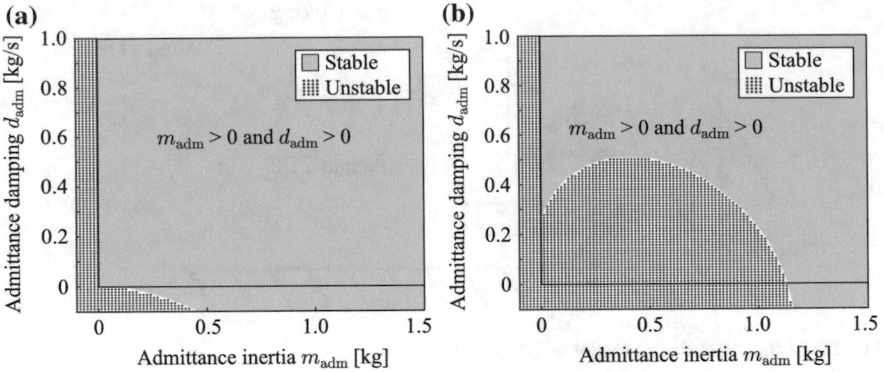

Fig. 5.3 Stable and unstable closed-loop eigenvalues of the linear, mobile robot with three DOF, depending on the admittance parameterization and the inertia couplings. In all cases, no Coriolis/centrifugal terms exist due to the use of prismatic joints ($\boldsymbol{C}_{qr} = \boldsymbol{0}$). **a** Without inertia couplings, $\boldsymbol{M}_{qr} = \boldsymbol{0}$. **b** With inertia couplings, $\boldsymbol{M}_{qr} \neq \boldsymbol{0}$

5.1.2 Control Design

The inertia and Coriolis/centrifugal couplings between admittance-controlled mobile base and torque-controlled upper body can destabilize the system. The compensation of these couplings is proposed in combination with the impedance controller in the operational space. Based on the insights from the analysis of the linear system, the inertia matrix ought to be decoupled such that it becomes symmetric again. Therefore, the impedance control law (5.12) is extended by an additional compensation term:

$$\begin{pmatrix} \boldsymbol{\tau}_{r,vir} \\ \boldsymbol{\tau}_{q} \end{pmatrix} = \boldsymbol{\tau}_{imp} + \boldsymbol{\tau}_{damp} + \boldsymbol{\tau}_{comp}, \tag{5.13}$$

$$\boldsymbol{\tau}_{comp} = \begin{pmatrix} \boldsymbol{0} \\ \boldsymbol{M}_{qr}\ddot{\boldsymbol{r}} + \boldsymbol{C}_{qr}\dot{\boldsymbol{r}} + \boldsymbol{g}_{q} \end{pmatrix}. \tag{5.14}$$

The compensation action $\boldsymbol{\tau}_{comp}$ is used to bring the dynamics into the standard form (2.6) of the rigid robot dynamics such that the resulting inertia matrix is symmetric and positive definite while (2.7) holds. The accelerations $\ddot{\boldsymbol{r}}$ do not have to be measured but they can be directly taken from the admittance simulation (5.1) due to $\dot{\boldsymbol{r}} \approx \dot{\boldsymbol{r}}_{des}$. Therefore, one can adopt

$$\ddot{\boldsymbol{r}} = \boldsymbol{M}_{adm}^{-1} \left(\boldsymbol{\tau}_{r,vir} + \boldsymbol{\tau}_{r,ext} - \boldsymbol{D}_{adm}\dot{\boldsymbol{r}}_{des} \right). \tag{5.15}$$

The controller implementation on the humanoid robot Rollin' Justin is illustrated in Fig. 5.4.

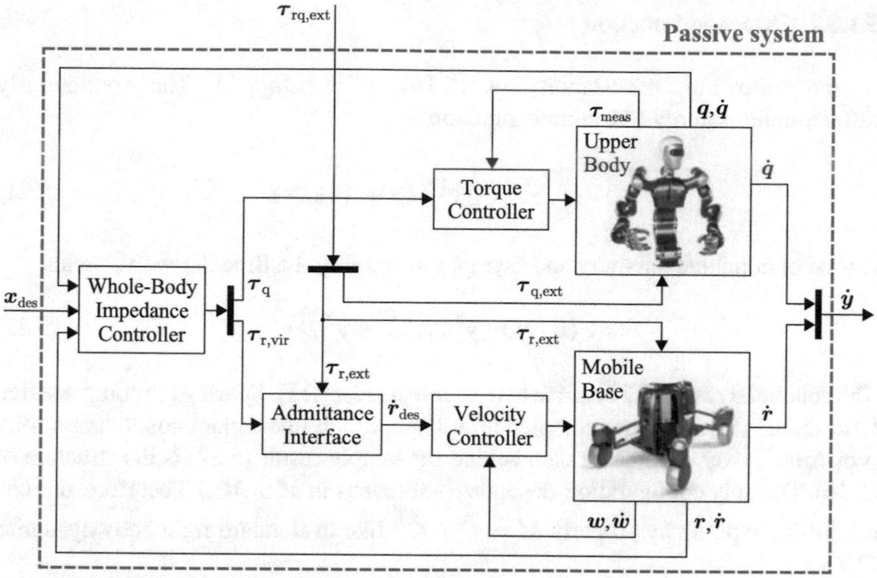

Fig. 5.4 Whole-body impedance control on the humanoid robot Rollin' Justin. The closed loop is passive w.r.t. the input $\tau_{rq,ext}$ and the output $\dot{y}$. The measured upper body joint torques are denoted by τ_{meas}

5.1.3 *Proof of Stability*

Now the dynamic equations take the form

$$\bar{M}\ddot{y} + \bar{C}\dot{y} + \bar{D}\dot{y} = \tau_{imp} + \tau_{rq,ext} \tag{5.16}$$

with

$$\bar{M} = \begin{pmatrix} M_{adm} & 0 \\ 0 & M_{qq} \end{pmatrix}, \tag{5.17}$$

$$\bar{C} = \begin{pmatrix} 0 & 0 \\ 0 & C_{qq} \end{pmatrix}, \tag{5.18}$$

$$\bar{D} = \begin{pmatrix} D_{adm} & 0 \\ 0 & D_{qq} \end{pmatrix}, \tag{5.19}$$

$$\tau_{rq,ext} = \begin{pmatrix} \tau_{r,ext} \\ \tau_{q,ext} \end{pmatrix}. \tag{5.20}$$

5.1.3.1 Storage Function

In the following, the stability of (5.16) is investigated. The continuously differentiable, energy-like storage function

$$V(\tilde{\boldsymbol{x}}, \dot{\boldsymbol{y}}) = \frac{1}{2}\dot{\boldsymbol{y}}^T \bar{\boldsymbol{M}} \dot{\boldsymbol{y}} + V_{\text{imp}}(\tilde{\boldsymbol{x}}) \tag{5.21}$$

is used to conclude passivity and asymptotic stability. Its time derivative yields

$$\dot{V}(\tilde{\boldsymbol{x}}, \dot{\boldsymbol{y}}) = \dot{\boldsymbol{y}}^T \boldsymbol{\tau}_{\text{rq,ext}} - \dot{\boldsymbol{y}}^T \bar{\boldsymbol{D}} \dot{\boldsymbol{y}}. \tag{5.22}$$

The beneficial result (5.22) is due to the control law (5.13). Sparing the compensation term (5.14) yields a more complex term from which one cannot conclude stability properties easily. Another reason behind the simple result (5.22) is the structure of (5.16). The only configuration-dependent submatrix in $\bar{\boldsymbol{M}}$ is $\boldsymbol{M}_{\text{qq}}$. Therefore, one can establish the passivity property $\dot{\bar{\boldsymbol{M}}} = \bar{\boldsymbol{C}} + \bar{\boldsymbol{C}}^T$ like in standard rigid body dynamics (2.7).

5.1.3.2 Passivity of the Closed Loop

Using the storage function $V(\tilde{\boldsymbol{x}}, \dot{\boldsymbol{y}})$ one can show strict output passivity of the closed loop w.r.t. the input $\boldsymbol{\tau}_{\text{rq,ext}}$ and the output $\dot{\boldsymbol{y}}$. That can be concluded from (5.22) for positive definite damping matrices $\bar{\boldsymbol{D}}$.

5.1.3.3 LaSalle's Invariance Principle

LaSalle's invariance principle can be applied for an undisturbed system (5.16), i.e. for $\boldsymbol{\tau}_{\text{rq,ext}} = \boldsymbol{0}$. Since $\dot{V}(\tilde{\boldsymbol{x}}, \dot{\boldsymbol{y}})$ is only negative semi-definite, the states satisfying $\dot{V}(\tilde{\boldsymbol{x}}, \dot{\boldsymbol{y}}) = 0$ have to be investigated. For $\dot{\boldsymbol{y}} = \ddot{\boldsymbol{y}} = \boldsymbol{0}$ and $\boldsymbol{\tau}_{\text{rq,ext}} = \boldsymbol{0}$, (5.16) delivers

$$\left(\frac{\partial V_{\text{imp}}(\tilde{\boldsymbol{x}}(\boldsymbol{y}))}{\partial \boldsymbol{y}}\right)^T = \boldsymbol{0}, \tag{5.23}$$

which only holds for $(\tilde{\boldsymbol{x}}, \dot{\boldsymbol{y}}) = (\boldsymbol{0}, \boldsymbol{0})$. Therefore, convergence to this equilibrium set can be concluded.

5.1.3.4 Robot Setup and Control Task Integration

With the insights from above, different robot setups and task integrations have to be distinguished: non-redundant robots, redundant robots where only damping is applied

Table 5.2 Controller parameters for the experiments on mobile impedance on Rollin' Justin

Gain	Value
$\boldsymbol{M}_{\text{adm}}$	diag(7.5 kg, 7.5 kg, 2.5 kg m^2)
$\boldsymbol{D}_{\text{adm}}$	diag(24 kg/s, 24 kg/s, 8 kg m^2/s)
$\boldsymbol{K}_{\text{tra}}$	diag(1000 N/m, 1000 N/m, 1000 N/m)
$\boldsymbol{K}_{\text{rot}}$	diag(100 Nm/rad, 100 Nm/rad, 100 Nm/rad)
$\boldsymbol{D}_{\text{qq}}$	$\mathcal{D}(\boldsymbol{M}(\boldsymbol{q}), \boldsymbol{K}_{\text{tra}}, \boldsymbol{K}_{\text{rot}}, \xi = 0.7)$

in the null space of the operational space task according to (5.19), and redundant systems where further tasks are applied in the null space of the operational space task.

Non-redundant robot: If $n_{\text{x}} = n_{\text{r}} + n_{\text{q}}$, the robot is non-redundant w.r.t. the operational space task and no null space exists. Asymptotic stability of the equilibrium can be shown in the configuration space for the equilibrium $(\boldsymbol{y}^*, \dot{\boldsymbol{y}})$ with $\tilde{\boldsymbol{x}}(\boldsymbol{y}^*) = \boldsymbol{0}$ and $\dot{\boldsymbol{y}} = \boldsymbol{0}$.

Redundant robot with null space damping: If $n_{\text{x}} < n_{\text{r}} + n_{\text{q}}$, the robot is redundant w.r.t. the operational space task, and a null space exists. If the control law (5.13) is applied, the overall damping matrix (5.19) is positive definite, i.e. the damping also covers the null space of the operational space task. Passivity and convergence to the equilibrium set $(\tilde{\boldsymbol{x}}(\boldsymbol{y}^*), \dot{\boldsymbol{y}}) = (\boldsymbol{0}, \boldsymbol{0})$ can be shown. The joint configuration $\boldsymbol{y}^*$ cannot be determined because the null space configuration is not unique.

Redundant robot with further null space tasks: If $n_{\text{x}} < n_{\text{r}} + n_{\text{q}}$, the robot is redundant w.r.t. the operational space task, and a null space exists. In order to properly define the null space behavior on position level, priority-based control concepts can be applied. The complete stability analysis of this case will be treated in Sect. 5.2. Asymptotic stability of the equilibrium $(\boldsymbol{y}^*, \dot{\boldsymbol{y}})$ can be shown that way, and $\boldsymbol{y}^*$ can be determined.

5.1.4 Experiments

The controller (5.13) is implemented and validated on Rollin' Justin as sketched in Fig. 5.4. The control gains are set according to Table 5.2, where the Cartesian stiffness matrix $\boldsymbol{K}$ is split up into its translational part $\boldsymbol{K}_{\text{tra}}$ and rotational part $\boldsymbol{K}_{\text{rot}}$. The upper body damping matrix $\boldsymbol{D}_{\text{qq}}$ is configuration-dependent and realizes damping ratios $\xi = 0.7$ w.r.t. the Cartesian impedance via the *Double Diagonalization* approach [ASOFH03]. The serial kinematic chain "platform[2]–torso–right arm" is considered in the following. Furthermore, an impedance in the torso is used to keep it within feasible regions in the body frame, the self-collision avoidance from Sect. 3.1 is

[2]In the following experiments, only the forward/backward motion of the platform is allowed in order to simplify the experimental evaluation.

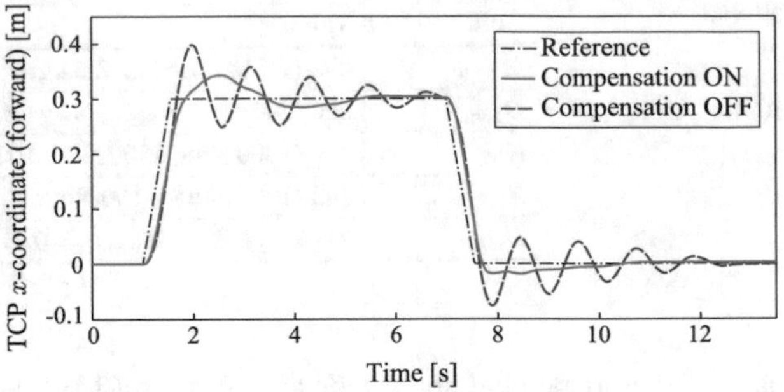

Fig. 5.5 Comparison of the TCP behavior in forward direction

applied, and a singularity avoidance for the arm is activated in the null space of the Cartesian impedance of the TCP to maximize the manipulability.

In the first experiment, a continuous forward-rest-backward trajectory of length 0.3 m is commanded to the right TCP. The transient spatial behavior of the TCP is shown in Fig. 5.5. With compensation, a nice impedance behavior is achieved with small overshooting. The overshoot is due to two reasons: First, the damping ratio for the Cartesian impedance is set to 0.7, cf. Table 5.2, which is a standard parameterization for this kind of lightweight robot [ASOFH03], that leads to fast response at the cost of small overshoots. Second, the kinematic velocity controller of the mobile base is not ideal. The introduced phase delay inevitably leads to slight uncertainties in the model such that the formulation (5.6) does not perfectly match the real dynamics anymore. Without compensation, the system oscillates significantly and takes a relatively long time to reach a steady state. However, by applying the chosen parameterization, the system still remains stable, even without compensation. If the admittance mass and damping is reduced, instability results without compensation of the inertia and Coriolis/centrifugal couplings. Such a scenario will be shown later.

In terms of the stability properties, the energy (5.21) is of high interest.[3] The top plot in Fig. 5.6 depicts the kinetic energies based on the admittance inertia, which shows strong oscillations without compensation of the inertia and Coriolis/centrifugal couplings. The large extent cannot be observed in the TCP deviation because Fig. 5.5 does not cover the null space motions. In contrast to that, the kinetic energy in Fig. 5.6 is influenced by the null space behavior. The potential energy in (5.21) is depicted in the center chart of Fig. 5.6. The sum of the kinetic energy and the potential energy is plotted in the bottom diagram. The stability problems without compensation are distinct. In the case of compensated inertia and Coriolis/centrifugal couplings, the total energy decreases quickly. The behavior on joint level is depicted in Fig. 5.7 at the example of the second torso joint. It shows the measured joint position, the joint velocity, the joint torque and the commanded compensation torque. One can

[3]The storage function (5.21) does not match the real physical energy due to the use of the admittance inertia instead of the real one. An overall potential energy including the other subtasks is not meaningful due to the null space projections as described in [DOAS13].

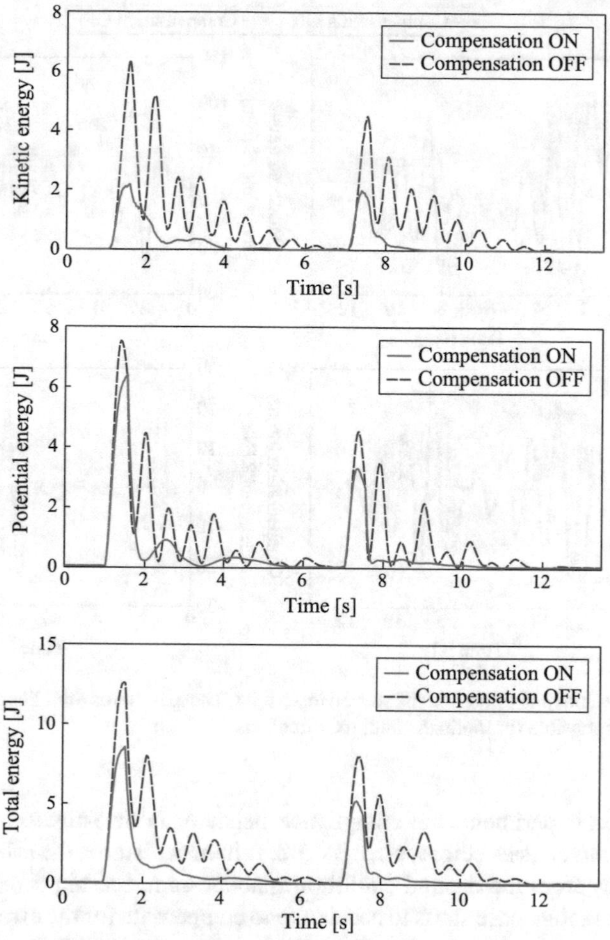

Fig. 5.6 Kinetic energy, potential energy, and the sum of both for scenarios with and without compensation. The kinetic energy is given by $\frac{1}{2}\dot{y}^T\bar{M}\dot{y}$ while the potential energy is defined by $V_{\text{imp}}(\bar{x})$

observe large differences in the motions depending on the compensation term. That effect becomes even more relevant if one bears in mind that this second torso axis is responsible for pitch motions and located directly above the mobile base. Thus, only small motions in this joint have a massive impact on the TCP pose due to the long lever arm. On joint torque level also large oscillations can be observed in the case of uncompensated inertia and Coriolis/centrifugal couplings. On the contrary, when these terms are compensated, only the inevitable peaks during the acceleration and deceleration phase are noteworthy in the plot. The bottom right chart in Fig. 5.7 shows the compensation torque in this specific joint.

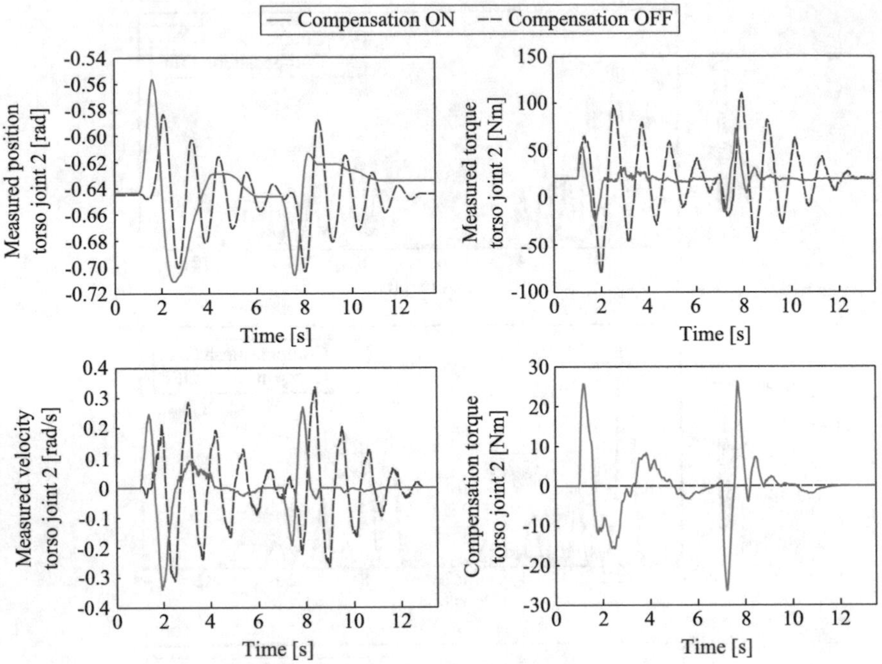

Fig. 5.7 Joint values and torques in the second torso joint. The axis of this joint lies in the horizontal plane and is responsible for motions about the pitch axis

In the next experiment, the compliance behavior is investigated for the same controller parameters as before, cf. Table 5.2. A human disturbs the robot by moving the TCP away from the desired equilibrium about 4 cm, see Fig. 5.8 (top). During that time, the mobile base starts to accelerate to compensate for the error, see Fig. 5.8 (center). When abruptly releasing the end-effector again, the TCP error converges properly in case of active compensation. If deactivated, massive oscillations can be observed in the error plots. When comparing the base position, one can clearly see that the platform changes its moving direction repeatedly, while it shows a proper behavior in case of active compensation. With compensation, the coordination between upper body and mobile base is reasonable: While the platform is still moving backwards ($3\,\text{s} < t < 8\,\text{s}$) from about $r_1 \approx 0.3\,\text{m}$, the TCP error is already very small with less than 1 cm. Hence, the null space is properly used to reconfigure the robot. Although the amplitude in the r_1-coordinate is smaller without compensation, one has to bear in mind that the platform alone weighs about 150 kg and changing the direction of motion twice per second demands a lot from the actuators and the power supply. The total energy (5.21) is plotted in Fig. 5.8 (bottom). The short-time increase in the

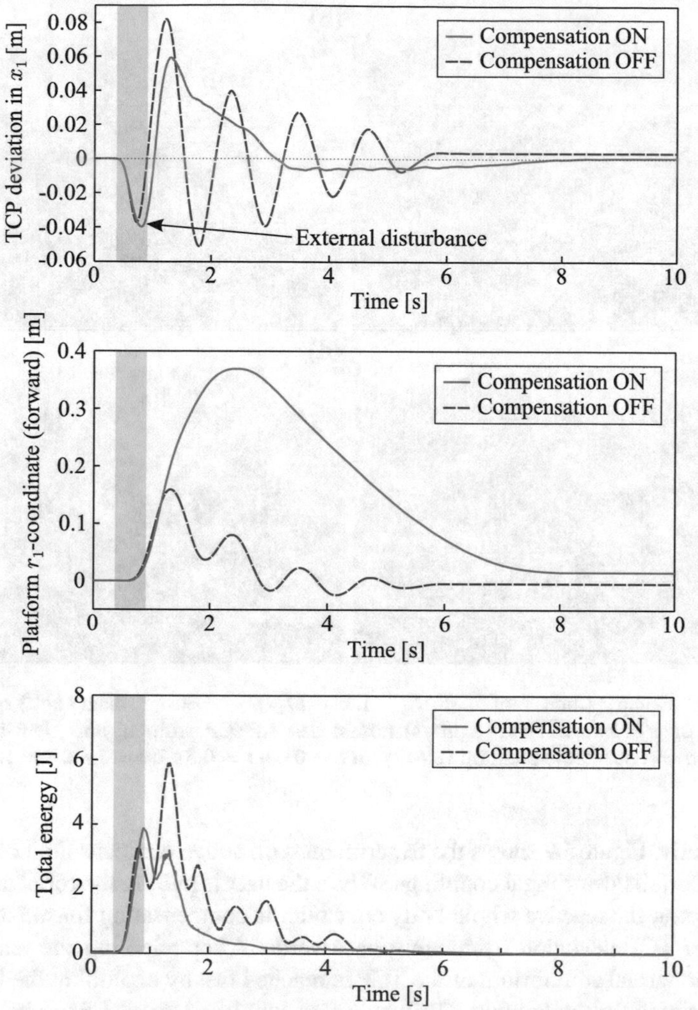

Fig. 5.8 Physical human-robot interaction: The robot is disturbed during the marked time period

energy with compensation ($1\,\text{s} < t < 1.5\,\text{s}$) can be traced back to the performance of the kinematic platform controller and the resulting model uncertainties in (5.16).

In the last experiment, a critical set of admittance parameters is applied such that instability occurs:

$$\boldsymbol{M}_{\text{adm}} = \text{diag}(3\,\text{kg}, 3\,\text{kg}, 1\,\text{kgm}^2), \;\; \boldsymbol{D}_{\text{adm}} = \text{diag}(21\,\text{kg/s}, 21\,\text{kg/s}, 7\,\text{kgm}^2/\text{s}).$$

Keeping the nominal platform mass of about 150 kg in mind, the velocity controller is instructed to reduce the perceived inertia to only 2 % of the original value. Moreover, the platform has to accelerate and decelerate the upper body of about 45 kg

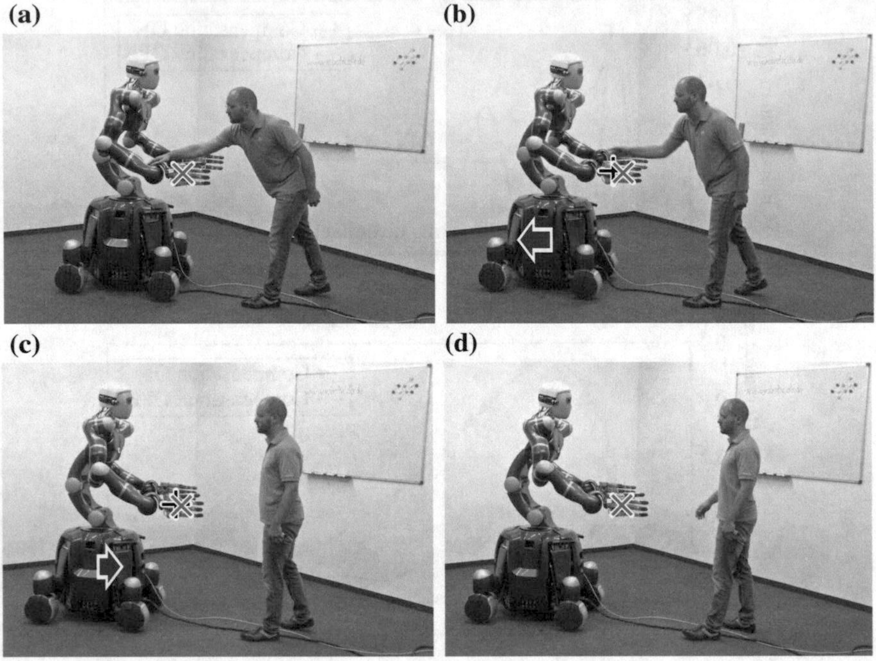

Fig. 5.9 Compensation of (5.14) for $\boldsymbol{M}_{\text{adm}} = \text{diag}(3\,\text{kg}, 3\,\text{kg}, 1\,\text{kgm}^2)$, $\boldsymbol{D}_{\text{adm}} = \text{diag}(21\,\text{kg/s}, 21\,\text{kg/s}, 7\,\text{kgm}^2/\text{s})$ (*cross*: desired TCP position, *plus sign*: actual TCP position, *arrow*: qualitative platform velocity). **a** $t = 0\,\text{s}$. **b** $t = 0.5\,\text{s}$. **c** $t = 1\,\text{s}$. **d** $t = 1.5\,\text{s}$

additionally. Figure 5.9 shows the experiment with active compensation of the inertia and Coriolis/centrifugal couplings. While the user is pulling the robot at the right end-effector, the reactive whole-body controller is compensating for the introduced Cartesian TCP deviation by moving backwards. After releasing the end-effector again, the virtual equilibrium of the TCP is reached fast by exploiting the kinematic redundancy in the upper body. The whole transient takes about 1.5 s only.

Figure 5.10 shows the same scenario without compensation of the dynamic couplings. Only slightly touching the end-effector immediately destabilizes the system. Notice that within only 1.5 s, i.e. between Fig. 5.10 (a) and (d), the platform moves a great distance forward (b) *and* backward (c), (d). At $t = 1.5$ s, the operator uses the emergency stop device. Due to the highly dynamic motion, the maximum permissible torque in the first horizontal torso axis (pitch motion) of 230 Nm is reached.

5.1.5 Discussion

The controller shapes the overall dynamics by modifying the inertia matrix and the Coriolis/centrifugal terms. Does such an extensive intervention cause problems

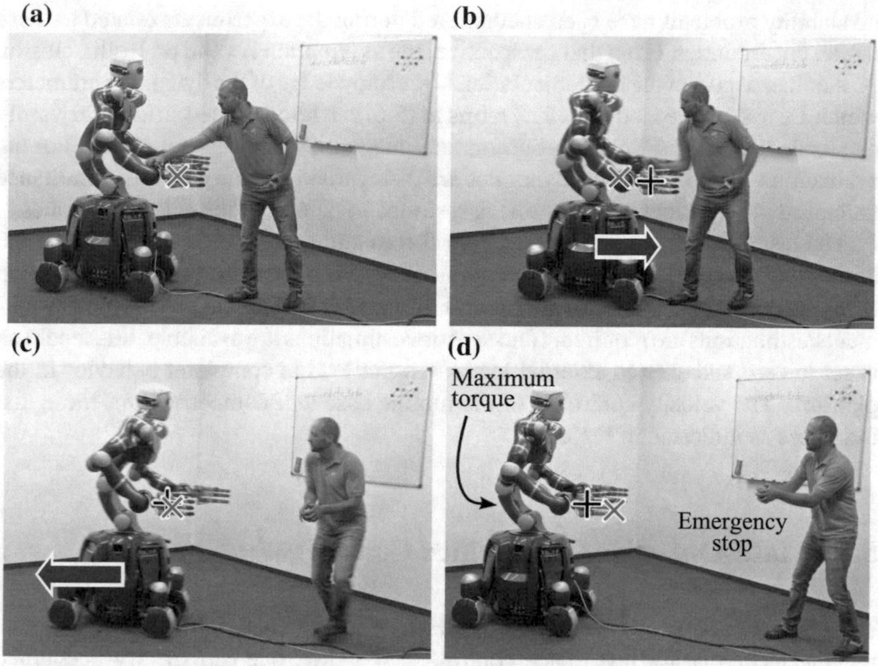

Fig. 5.10 No compensation (5.14) for $\boldsymbol{M}_{\text{adm}} = \text{diag}(3\,\text{kg}, 3\,\text{kg}, 1\,\text{kgm}^2)$, $\boldsymbol{D}_{\text{adm}} = \text{diag}(21\,\text{kg/s}, 21\,\text{kg/s}, 7\,\text{kgm}^2/\text{s})$ (*cross*: desired TCP position, *plus sign*: actual TCP position, *arrow*: qualitative platform velocity). **a** $t = 0\,\text{s}$. **b** $t = 0.5\,\text{s}$. **c** $t = 1\,\text{s}$. **d** $t = 1.5\,\text{s}$

in terms of robustness and availability of measurements? The control law requires model-based calculations to cancel elements in the Coriolis/centrifugal matrix but only positions and velocities are used in the feedback law. These signals are usually measured or the velocities are derived by differentiating the positions w.r.t. time without jeopardizing the robustness. The same applies to active damping (5.11). Modifying the inertia matrix requires the availability of acceleration measurements. Here it is advantageous that the respective term in (5.14) only requires the Cartesian base accelerations which can be directly taken from the admittance simulation (5.1) without referring to additional measurements because of the assumption $\dot{\boldsymbol{r}} \approx \dot{\boldsymbol{r}}_{\text{des}}$. If $\boldsymbol{M}_{\text{qr}}$ was not cancelled and $\boldsymbol{M}_{\text{qr}}^T$ was inserted in the upper right element of $\bar{\boldsymbol{M}}$, one would also obtain a symmetric inertia matrix and stability can be shown. However, then the admittance would not feature the desired behavior (5.1), and feedback of the upper body joint accelerations would be required in the control law.

In the stability analysis, it has been assumed that the platform velocity controller compensates for any disturbances (cf. Fig. 5.1). The experiments on Rollin' Justin revealed that the velocity control [GFASH09] of the platform does not perform as desired due to phase delays and amplitude errors. This inaccuracy leads to differences between the computed dynamic model and the actual system dynamics. However, no stability problems have been encountered during the experiments related to these model uncertainties. From that perspective, the assumption is valid on Rollin' Justin.

Another aspect of the implementation is the knowledge of the dynamic parameters which are used in the feedback. The terms in (5.6) can be computed straightforwardly by standard symbolic algebra programs and integrated in the real-time code. For the evaluation of (5.6), the wheel dynamics are disregarded and the platform admittance (achieved via the kinematic control) is assumed to be part of the actual dynamics.

The last aspect addresses the external forces and torques exerted on the mobile base. In (5.1), they are used in the admittance simulation. To provide interaction compliance also w.r.t. external forces and torques exerted on the platform, measurements/estimations are required. If no sensors/estimations are available, this feedback is set to zero and exerted external loads will not lead to compliant behavior in the platform. The velocity controller of the mobile base will compensate for these disturbances as indicated in Fig. 5.1.

5.2 Multi-Objective Compliance Control

The *Operational Space Formulation* (OSF) by Khatib [Kha87] is probably the most popular technique for task space control. In the OSF, one can specify decoupled linear dynamics on the main task level, which is similar to the feedback linearization approach in nonlinear control theory. That enables to separately handle different simultaneous objectives. A dynamically consistent hierarchy among these tasks can be arranged by utilizing null space projections as described in Chap. 4. However, a proof of stability for the complete robot using the OSF is not known [NCM+08]. Early works on redundancy resolution involved the augmentation of additional task coordinates [Bai85], but that led to new algorithmic singularities[4] and inertial couplings. The extended task space approach was generalized to compliant motion control by Peng and Adachi in [PA93].

If external forces act on the robot, it is necessary to consider them in the controller to get a decoupled behavior also w.r.t. these loads. While forces and torques exerted on the end-effector can usually be measured by a six-axis force-torque sensor, external loads acting in the null space are more problematic and require observers or additional instrumentation. Compliance controllers can be implemented without measurement of external forces if the desired impedance is characterized by a desired compliance in terms of stiffness and damping. In this case the desired inertia corresponds to the natural inertia of the robot. Natale et al. [NSV99] extended a spatial Cartesian impedance controller by an additional null space control action and showed asymptotic convergence of a null space velocity error term. In the context of impedance control, the minimization of the quadratic norm of a lower-priority impedance error in a two-level hierarchy has been treated in [PAW10]. Multi-objective whole-body

[4] Algorithmic singularities arise when lower-priority tasks and higher-priority tasks conflict with each other, i.e. stacking their Jacobian matrices results in a matrix that does not have full row-rank.

compliance control concepts utilizing joint torque sensing have recently been developed [DWASH12b, DWASH12a]. Several works in the literature also address prioritized multi-task control at the kinematic level. Chiaverini [Chi97] proposed a singularity-robust inverse kinematics for a simple hierarchy of two tasks. Antonelli [Ant09] provided a proof of stability for prioritized closed-loop inverse kinematics. However, it was limited to the kinematic case. Another issue in the context of hierarchy-based control is the occurrence of discontinuities in the control law due to a change in the rank of the Jacobian matrices or in the inequality constraints. Mansard et al. addressed these issues in [MKK09]. In [OKN08], a stability analysis for a null space compliance controller with a simple hierarchy of two priority levels has been presented. A primary Cartesian task is inertially decoupled from a null space task by proper choice of coordinates, and asymptotic stability is shown utilizing semi-definite Lyapunov functions. The dynamics formulation is based on [Par99], where the primary task coordinates are augmented by appropriate dynamically consistent null space velocities. As a result, the inertia matrix of the error dynamics becomes block-diagonal which corresponds to the decoupling of the kinetic energies on the involved priority levels. A similar dynamics relation was also utilized by Oh [OCY98] for the implementation of an impedance controller. The stability analysis, however, was limited to null space damping. The non-integrability of the null space velocities represents a major obstacle for such a stability analysis [OKN08].

Nakanishi et al. compared eight established OSF controllers from a theoretical and empirical perspective [NCM+08]. Although exponential stability can be shown for the main task, the authors state that "*null space dynamics so far resist insightful general analytical investigations (...) If stability could be proven for this family of operational space controllers, operational space control would be lifted to a more solid foundation*". Control approaches for the tracking problem of redundant manipulators have been proposed lately, which also enable a stability analysis of the null space error. The subtask convergence is handled by a kinematic approach and the resulting joint velocity is utilized within a task space controller such as [ZDWS04, SKVS12]. An extended OSF approach with stable null space posture control was presented in [SPP13]. However, the limitation to posture optimization in the null space limits the potential of that approach.

In this section, the work initiated in [OKN08] is extended to hierarchies with an arbitrary number of priority levels [DOAS13, ODAS15] as illustrated in Fig. 5.11. In [OKN08], a null space compliance controller was developed to decouple a high-priority task from a single null space task. The first contribution of the concept presented here is the derivation of a dynamics formulation which features a hierarchical decoupling between all tasks. Based on this formulation, a compliance controller for all priority levels is implemented which does *not* require feedback of external forces or inertia shaping. The second contribution is the stability analysis of the closed-loop system via semi-definite Lyapunov functions. The analysis is based on passivity theory and is made possible thanks to the specific dynamics representation. It is the first proof of stability for a generic task hierarchy with an arbitrary number of priority levels.

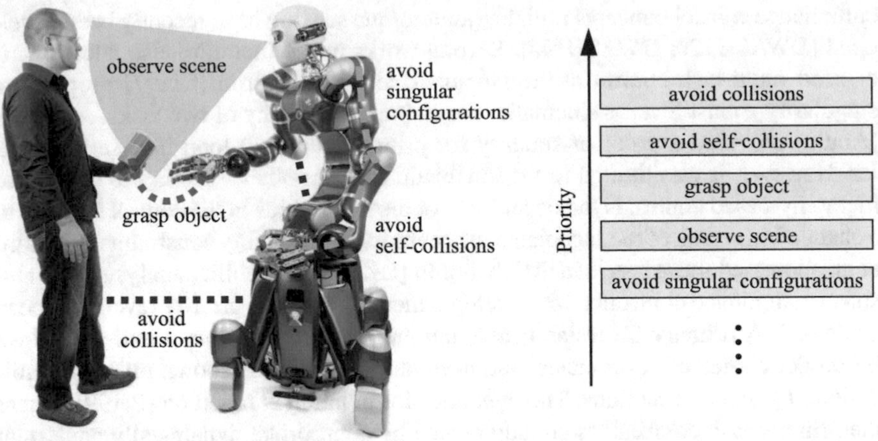

Fig. 5.11 Example of a task hierarchy with several objectives (*left*). A priority list (*right*) weights the importance of the involved tasks

After a short introduction to the problem, the basic features of the approach by Ott et al. [OKN08] are recapitulated in Sect. 5.2.2.1. Afterwards, the concept is extended to the general case of a hierarchy with an arbitrary number of priority levels [DOAS13, ODAS15] in Sect. 5.2.2.2. The control design is presented in Sect. 5.2.3. The stability analysis is given in Sect. 5.2.4. The chapter closes with a validation by simulations and experiments in Sect. 5.2.5 and the discussion on the stability properties in Sect. 5.2.6.

5.2.1 Problem Formulation

5.2.1.1 Definition of Multi-Objective Compliance Control

Based on the rigid body dynamics (2.6) for a generic n-DOF robot, a hierarchical dynamics representation is derived. The total number of r task coordinate vectors is defined by the mappings

$$\boldsymbol{x}_i = \boldsymbol{f}_i(\boldsymbol{q}) \in \mathbb{R}^{m_i} \ \forall i, \ \ 1 \leq i \leq r \tag{5.24}$$

These r tasks with the respective dimensions m_i are arranged in a hierarchical order. The priority levels are defined such that $i = 1$ is top priority and $i_a < i_b$ implies that i_a is located higher in the priority list than i_b. The mappings from joint velocities to task velocities are given by the Jacobian matrices $\boldsymbol{J}_i(\boldsymbol{q}) \in \mathbb{R}^{m_i \times n} \ \forall i, \ 1 \leq i \leq r$:

$$\dot{\boldsymbol{x}}_i = \boldsymbol{J}_i(\boldsymbol{q})\dot{\boldsymbol{q}}, \quad \boldsymbol{J}_i(\boldsymbol{q}) = \frac{\partial \boldsymbol{f}_i(\boldsymbol{q})}{\partial \boldsymbol{q}}. \tag{5.25}$$

In the following analysis, all Jacobian matrices are assumed to be non-singular and consequently of full row rank. The primary task $i = 1$ has the dimension $m_1 < n$ so that a kinematic redundancy of $n - m_1$ DOF remains to accomplish subtasks in its null space. The goal is a prioritized compliance control with a task hierarchy that complies with the following conditions:

1. A task with lower priority i_b may not disturb any task with higher priority i_a, where $i_a < i_b$. A low-priority task is executed in the null space of *all* higher-priority tasks.
2. A compliance control task can be described by a positive definite potential function $V_i(\tilde{\boldsymbol{x}}_i(\boldsymbol{q}))$ related to the task coordinate $\tilde{\boldsymbol{x}}_i(\boldsymbol{q}) = \boldsymbol{x}_i(\boldsymbol{q}) - \boldsymbol{x}_{i,\text{des}}$ with the virtual equilibrium $\boldsymbol{x}_{i,\text{des}}$. The damping is specified by a positive definite damping matrix[5] $\boldsymbol{D}_i \in \mathbb{R}^{m_i \times m_i}$.
3. The dimension m_r of the lowest-level task may be larger than $n - \sum_{i=1}^{r-1} m_i$ such that the dimension n of the joint space is exceeded by the total dimension $\sum_{i=1}^{r} m_i$ of all tasks. A typical example of the lowest-level task is a joint level compliance.

5.2.1.2 Relation to the Operational Space Formulation

Prior to the presentation of the approach, several important properties of the OSF [Kha87] are reviewed for which an extension to multiple, prioritized tasks exists [SK05]. The discussion is mainly limited to the issue of inertia shaping related to the main task. A more detailed analysis of different operational space concepts in kinematically redundant manipulators is provided in [NCM+08]. The control goal in the OSF is to obtain decoupled dynamics in the main task space $\boldsymbol{x}_1$. In case of interaction control, i.e. in the presence of external loads, the desired dynamics takes the form

$$\boldsymbol{\Lambda}_{1,\text{des}}\ddot{\tilde{\boldsymbol{x}}}_1 + \boldsymbol{D}_1\dot{\tilde{\boldsymbol{x}}}_1 + \boldsymbol{K}_1\tilde{\boldsymbol{x}}_1 = \boldsymbol{F}_{1,\text{ext}} \tag{5.26}$$

where $\boldsymbol{K}_1 \in \mathbb{R}^{m_1 \times m_1}$ and $\boldsymbol{D}_1 \in \mathbb{R}^{m_1 \times m_1}$ are the stiffness and damping matrices, respectively, and $\boldsymbol{\Lambda}_{1,\text{des}} \in \mathbb{R}^{m_1 \times m_1}$ denotes the inertia matrix. Note that all of the three matrices can be specified as desired, but $\boldsymbol{\Lambda}_{1,\text{des}}$ gets the additional subscript "des" because $\boldsymbol{\Lambda}_1(\boldsymbol{q}) = (\boldsymbol{J}_1(\boldsymbol{q})\boldsymbol{M}(\boldsymbol{q})^{-1}\boldsymbol{J}_1(\boldsymbol{q})^T)^{-1}$ is already reserved for the unmodified (natural) operational space inertia in this chapter. The external force $\boldsymbol{F}_{1,\text{ext}} \in \mathbb{R}^{m_1}$ in (5.26) is related to the main task and collocated to $\dot{\boldsymbol{x}}_1$. For positioning tasks one usually chooses $\boldsymbol{\Lambda}_{1,\text{des}} = \boldsymbol{I}$ and removes the external force from the right hand side. It can easily be verified that the desired dynamics (5.26) can be exactly achieved by the feedback linearization

[5] Depending on the application, the damping matrix can be chosen constant or configuration-dependent as long as it is a positive definite matrix, see e.g. [ASOFH03].

$$\boldsymbol{\tau} = \boldsymbol{C}(\boldsymbol{q}, \dot{\boldsymbol{q}})\dot{\boldsymbol{q}} + \boldsymbol{g}(\boldsymbol{q}) + \boldsymbol{J}_1(\boldsymbol{q})^T \boldsymbol{F}_1 + \boldsymbol{N}_2(\boldsymbol{q})\boldsymbol{\tau}_2, \tag{5.27}$$

$$\begin{aligned}\boldsymbol{F}_1 = &-\boldsymbol{\Lambda}_1(\boldsymbol{q})\boldsymbol{\Lambda}_{1,\mathrm{des}}^{-1}(\boldsymbol{D}_1\dot{\boldsymbol{x}}_1 + \boldsymbol{K}_1\tilde{\boldsymbol{x}}_1) - \boldsymbol{\Lambda}_1(\boldsymbol{q})\dot{\boldsymbol{J}}_1(\boldsymbol{q}, \dot{\boldsymbol{q}})\dot{\boldsymbol{q}} \\ &+ (\boldsymbol{\Lambda}_1(\boldsymbol{q})\boldsymbol{\Lambda}_{1,\mathrm{des}}^{-1} - \boldsymbol{I})\boldsymbol{F}_{1,\mathrm{ext}},\end{aligned} \tag{5.28}$$

where $\boldsymbol{N}_2(\boldsymbol{q})\boldsymbol{\tau}_2$ is a dynamically consistent null space torque according to Sect. 4.2.2 which does not affect the main task dynamics (5.26). The control law uses feedback of external forces. However, in the special case where $\boldsymbol{\Lambda}_{1,\mathrm{des}} = \boldsymbol{\Lambda}_1(\boldsymbol{q})$, no feedback of $\boldsymbol{F}_{1,\mathrm{ext}}$ is required. In consequence of this compliance control using the natural inertia, the closed-loop dynamics becomes nonlinear. The same issue related to force feedback appears in the lower-priority tasks, which are implemented in the null space of $\boldsymbol{J}_1(\boldsymbol{q})$ via $\boldsymbol{N}_2(\boldsymbol{q})\boldsymbol{\tau}_2$. If $\boldsymbol{\Lambda}_{1,\mathrm{des}} \neq \boldsymbol{\Lambda}_1(\boldsymbol{q})$, the generalized external forces related to the subtasks need to be measured as well, which implies that a measurement of the whole term $\boldsymbol{\tau}_{\mathrm{ext}}$ becomes necessary. While external forces and torques exerted on the end-effector can usually be measured by a force-torque sensor at the tip, the measurement of all joint torques in a redundant robot is often not directly available and must be implemented via additional observers. Moreover, as highlighted in [NCM+08], due to the projection in the respective null spaces, the task dynamics of the lower tasks cannot be analyzed independently from each other which poses an obstacle for the stability analysis that has not been overcome yet.

5.2.2 *Hierarchical Dynamics Representation*

The control approach to be presented later does not require the measurement of the generalized external forces due to avoidance of inertia shaping by $\boldsymbol{\Lambda}_{1,\mathrm{des}} = \boldsymbol{\Lambda}_1(\boldsymbol{q})$. It utilizes a dynamics formulation in a new set of velocity coordinates, where the dynamics of each hierarchy level are largely decoupled from all other levels. While a separation of the inertial terms can be achieved by a pure change of coordinates, the decoupling of centrifugal and Coriolis effects has to be performed by active control (Sect. 5.2.3). First, the special case of only one null space task [OKN08] is recapitulated. Second, the generalization to multiple null space subtasks is presented as in [DOAS13]. Dynamic consistency is preserved in both cases.

5.2.2.1 Hierarchy with Two Priority Levels

One can reformulate the rigid body dynamics (2.6) to decouple the main task dynamics from the null space dynamics [OKN08]. In [Bai85], Baillieul proposed additional task coordinates to describe the complete dynamics. However, that choice leads to new algorithmic singularities [Chi97]. Park et al. introduced $n - m_1$ null space velocity coordinates $\boldsymbol{v}_2 = \bar{\boldsymbol{J}}_2(\boldsymbol{q})\dot{\boldsymbol{q}}$ in [PCY99]. This approach is followed and adapted here. The matrix $\bar{\boldsymbol{J}}_2(\boldsymbol{q})$ has to be chosen in a way such that the so-called *extended* Jacobian matrix $\bar{\boldsymbol{J}}(\boldsymbol{q}) \in \mathbb{R}^{n \times n}$, defined by

$$\begin{pmatrix} v_1 \\ v_2 \end{pmatrix} = \bar{J}(q)\dot{q} = \begin{pmatrix} \bar{J}_1(q) \\ \bar{J}_2(q) \end{pmatrix} \dot{q}, \tag{5.29}$$

is non-singular. For consistency in the notations, $v_1 \in \mathbb{R}^{m_1}$ and $\bar{J}_1(q) \in \mathbb{R}^{m_1 \times n}$ have been introduced with $v_1 = \dot{x}_1$ and $\bar{J}_1(q) = J_1(q)$. In general, the null space velocity v_2 is non-integrable. In other words, compatible null space coordinates $n_2(q)$ do *not* exist such that $\bar{J}_2(q) = \partial n_2(q)/\partial q$ would hold. That fact is an obstacle for designing null space compliance controllers [OKN08]. In the analysis in Sect. 5.2.4, it will be shown that the issue can be overcome by means of a theorem on semi-definite Lyapunov functions. A dynamically consistent null space projection according to Definition 4.2 is achieved by

$$\bar{J}_2(q) = \left(Z_2(q)M(q)Z_2(q)^T\right)^{-1} Z_2(q)M(q) \tag{5.30}$$

$$= \left(Z_2(q)^{M^{-1}+}\right)^T, \tag{5.31}$$

where $Z_2(q) \in \mathbb{R}^{m_2 \times n}$ is a full-row-rank base of the null space of $J_1(q)$ that complies with the condition $J_1(q)Z_2(q)^T = \mathbf{0}$. For more details on the derivation, see [OKN08]. The singular value decomposition (SVD) is one numerical method to compute such a matrix $Z_2(q)$. But there also exist several analytical approaches such as [HV91, CW93]. Since, by assumption, $\bar{J}_1(q)$ also has full row rank, $\bar{J}(q)$ is invertible[6] and its inverse is given by

$$\bar{J}(q)^{-1} = \left(J_1(q)^{M+},\ Z_2(q)^T\right) = \left(\bar{J}_1(q)^{M+},\ \bar{J}_2(q)^{M+}\right), \tag{5.32}$$

where $J_1(q)^{M+}$ is the dynamically consistent inverse (cf. Sect. 4.2.2) given by

$$J_1(q)^{M+} = M(q)^{-1} J_1(q)^T \left(J_1(q)M(q)^{-1}J_1(q)^T\right)^{-1}. \tag{5.33}$$

With (5.29) and (5.32), the joint velocity $\dot{q}$ can be expressed as a function of v_1 and v_2:

$$\dot{q} = \bar{J}_1(q)^{M+} v_1 + \bar{J}_2(q)^{M+} v_2. \tag{5.34}$$

By applying that coordinate transformation (on the basis of (2.6)), the dynamics can be reformulated as

$$\Lambda(q) \begin{pmatrix} \dot{v}_1 \\ \dot{v}_2 \end{pmatrix} + \mu(q, \dot{q}) \begin{pmatrix} v_1 \\ v_2 \end{pmatrix} + \bar{J}(q)^{-T} g(q) = \bar{J}(q)^{-T} \left(\tau + \tau_{\text{ext}}\right). \tag{5.35}$$

The decoupled (i.e. block-diagonal) inertia matrix $\Lambda(q) \in \mathbb{R}^{n \times n}$ and the Coriolis/centrifugal matrix $\mu(q, \dot{q}) \in \mathbb{R}^{n \times n}$ are given by

[6] The proof can be found in [Ott08].

$$\boldsymbol{\Lambda}(\boldsymbol{q}) = \bar{\boldsymbol{J}}(\boldsymbol{q})^{-T}\boldsymbol{M}(\boldsymbol{q})\bar{\boldsymbol{J}}(\boldsymbol{q})^{-1} = \mathrm{diag}(\boldsymbol{\Lambda}_1(\boldsymbol{q}), \boldsymbol{\Lambda}_2(\boldsymbol{q})), \tag{5.36}$$

$$\boldsymbol{\mu}(\boldsymbol{q}, \dot{\boldsymbol{q}}) = \boldsymbol{\Lambda}(\boldsymbol{q})\left(\bar{\boldsymbol{J}}(\boldsymbol{q})\boldsymbol{M}(\boldsymbol{q})^{-1}\boldsymbol{C}(\boldsymbol{q}, \dot{\boldsymbol{q}}) - \dot{\bar{\boldsymbol{J}}}(\boldsymbol{q}, \dot{\boldsymbol{q}})\right)\bar{\boldsymbol{J}}(\boldsymbol{q})^{-1}, \tag{5.37}$$

with

$$\boldsymbol{\Lambda}_1(\boldsymbol{q}) = \left(\boldsymbol{J}_1(\boldsymbol{q})\boldsymbol{M}(\boldsymbol{q})^{-1}\boldsymbol{J}_1(\boldsymbol{q})^T\right)^{-1}, \tag{5.38}$$

$$\boldsymbol{\Lambda}_2(\boldsymbol{q}) = \boldsymbol{Z}_2(\boldsymbol{q})\boldsymbol{M}(\boldsymbol{q})\boldsymbol{Z}_2(\boldsymbol{q})^T. \tag{5.39}$$

Notice that the block-diagonal structure of $\boldsymbol{\Lambda}(\boldsymbol{q})$ is a result of utilizing the inertia matrix when computing $\bar{\boldsymbol{J}}_2(\boldsymbol{q})$ in (5.31) [PCY99]. This specific form of the dynamics formulation for a redundant manipulator using inertially decoupled null space velocity coordinates is useful for the design of feedback controllers. While such a decoupling can also be achieved by feedback linearization, e.g. in the OSF in Sect. 5.2.1.2, the controller design in Sect. 5.2.3 aims at a passivity-based compliance controller which deliberately avoids inertia shaping so that it can be implemented without measurement of external forces acting on the robot.

5.2.2.2 Hierarchy with an Arbitrary Number of Priority Levels

In the preceding section, the whole null space of the primary task has been exploited for the definition of the null space velocities $\boldsymbol{v}_2$. That procedure from [OKN08] is now extended to an arbitrary number of priority levels according to

$$\boldsymbol{v}_i = \bar{\boldsymbol{J}}_i(\boldsymbol{q})\dot{\boldsymbol{q}} \;\; \forall i,\; 1 \leq i \leq r \tag{5.40}$$

with the invertible[7] extended Jacobian matrix

$$\begin{pmatrix} \boldsymbol{v}_1 \\ \vdots \\ \boldsymbol{v}_r \end{pmatrix} = \bar{\boldsymbol{J}}(\boldsymbol{q})\dot{\boldsymbol{q}} = \begin{pmatrix} \bar{\boldsymbol{J}}_1(\boldsymbol{q}) \\ \vdots \\ \bar{\boldsymbol{J}}_r(\boldsymbol{q}) \end{pmatrix}\dot{\boldsymbol{q}}. \tag{5.41}$$

The question arises how to partition the null space of $\boldsymbol{J}_1(\boldsymbol{q})$ for the remaining subtasks. In order to get a similar structure as (5.35) with a block-diagonal inertia matrix $\boldsymbol{\Lambda}(\boldsymbol{q})$, the respective null space base matrices $\boldsymbol{Z}_i(\boldsymbol{q}) \in \mathbb{R}^{m_i \times n}$ $\forall i,\; 2 \leq i \leq r$ and their related null space Jacobian matrices have to be found such that the (equivalent) identities

$$\bar{\boldsymbol{J}}_i(\boldsymbol{q})\boldsymbol{M}(\boldsymbol{q})^{-1}\bar{\boldsymbol{J}}_j(\boldsymbol{q})^T = \boldsymbol{0}, \tag{5.42}$$

$$\boldsymbol{Z}_i(\boldsymbol{q})\boldsymbol{M}(\boldsymbol{q})\boldsymbol{Z}_j(\boldsymbol{q})^T = \boldsymbol{0} \tag{5.43}$$

[7] A further discussion on this requirement will be given in Sect. 5.2.6. The proof of invertibility of $\bar{\boldsymbol{J}}$ is provided in Appendix C.2.

Table 5.3 Matrix dimensions in the SVD

Matrix	Dimensions
$U_{i-1}(q)$	$\sum_{j=1}^{i-1} m_j \times \sum_{j=1}^{i-1} m_j$
$S_{i-1}(q)$	$\sum_{j=1}^{i-1} m_j \times n$
$V_{i-1}(q)$	$n \times n$
$\Sigma_{i-1}(q)$	$\sum_{j=1}^{i-1} m_j \times \sum_{j=1}^{i-1} m_j$
$X_{i-1}(q)$	$\sum_{j=1}^{i-1} m_j \times n$
$Y_{i-1}(q)$	$(n - \sum_{j=1}^{i-1} m_j) \times n$

hold for $i \neq j$. Equation (5.42) states that $\bar{J}_i(q)$ and $\bar{J}_j(q)$ are orthogonal w.r.t. the metric tensor $M(q)^{-1}$ for $i \neq j$, while (5.43) says that $Z_i(q)$ and $Z_j(q)$ are orthogonal w.r.t. the metric tensor $M(q)$ for $i \neq j$. These conditions for dynamic decoupling are achieved by

$$\begin{aligned}\bar{J}_i(q) &= \left(Z_i(q)M(q)Z_i(q)^T\right)^{-1} Z_i(q)M(q) \\ &= (Z_i(q)^{M^{-1}+})^T \qquad (5.44)\end{aligned}$$

for $i = 1 \ldots r$. In the following, an expression for $Z_i(q)$ is derived which generates the same control action for each task force as the dynamically consistent projector $N_i(q) \in \mathbb{R}^{n \times n}$ from Sect. 4.2.2, which fulfills Definition 4.2 thanks to the generalized weighting matrix (4.33).[8] On each hierarchy level, the classical projector is given by

$$\begin{aligned}N_i(q) &= I - J_{i-1}^{\text{aug}}(q)^T \left(\left(J_{i-1}^{\text{aug}}(q)\right)^{M+}\right)^T \\ &= M(q)Y_{i-1}(q)^T \left(Y_{i-1}(q)M(q)Y_{i-1}(q)^T\right)^{-1} Y_{i-1}(q), \qquad (5.45)\end{aligned}$$

with the augmented Jacobian matrix introduced in (4.10) that takes all tasks "down" to level $i-1$ into account. The null space base $Y_{i-1}(q)$ is obtained via SVD of $J_{i-1}^{\text{aug}}(q)$:

$$J_{i-1}^{\text{aug}}(q) = U_{i-1}(q)S_{i-1}(q)V_{i-1}(q)^T, \qquad (5.46)$$
$$V_{i-1}(q) = \left(X_{i-1}(q)^T, Y_{i-1}(q)^T\right), \qquad (5.47)$$
$$S_{i-1}(q) = (\Sigma_{i-1}(q), 0). \qquad (5.48)$$

All dimensions are listed in Table 5.3. The matrices $U_{i-1}(q)$ and $V_{i-1}(q)$ are orthonormal, the rectangular diagonal matrix $S_{i-1}(q)$ contains the $\sum_{j=1}^{i-1} m_j$ singular values of $J_{i-1}^{\text{aug}}(q)$ in its submatrix $\Sigma_{i-1}(q)$. The range space of the augmented Jacobian matrix is defined by $X_{i-1}(q)$ while its null space is determined by $Y_{i-1}(q)$. The null space base matrices $Z_i(q)$ and the null space Jacobian matrices $\bar{J}_i(q)$ must fulfill

[8] To simplify the notations, the corresponding matrices are chosen according to the solution (4.29), i.e. $B_X = M(q)$ and $B_Y = I$, without loss of generality.

$$J_{i-1}^{\text{aug}}(q)Z_i(q)^T = \mathbf{0}, \tag{5.49}$$

but they should only span the subspace of $N_i(q)$ which can actually be used to execute the subtask on level i. Since the null space decomposition is not unique, an additional condition must be imposed: A task force F_i from level i mapped via $\bar{J}_i(q)^T Z_i(q) J_i(q)^T$ has to generate the same control action τ_i^{p} as if mapped via $N_i(q)J_i(q)^T$, i.e.

$$\bar{J}_i(q)^T Z_i(q) J_i(q)^T = N_i(q) J_i(q)^T. \tag{5.50}$$

Equation (5.50) closes the gap to the classical approach [Kha87]. An interpretation of the left side is the successive mapping of a task force starting in the original space (5.25), see Fig. 5.12. The task force is initially mapped onto the joint torque space via $J_i(q)^T$ as in the standard impedance control framework (Sect. 2.2.1). Afterwards, the resultant joint torque is mapped onto a (dynamically consistent) null space task force space, which is defined by the new, local null space directions $Z_i(q)$. In the third step, the obtained dynamically consistent task force is mapped back onto the joint torque space by the corresponding transposed Jacobian matrix $\bar{J}_i(q)^T$.

The solution for $Z_i(q)$ is

$$Z_i(q) = \begin{cases} (J_1(q)^{M+})^T & \text{if } i = 1 \\ J_i(q)M(q)^{-1}N_i(q) & \text{if } 1 < i < r \\ Y_{r-1}(q) & \text{if } i = r \end{cases}. \tag{5.51}$$

The proof of (5.50) for (5.51) with (5.44) can be found in Appendix C.1. The presence of $J_i(q)$ (for $1 < i < r$) is justified by the necessity of reducing the available null space to the subspace which is really used on the considered level. That reduction is actually given by $J_i(q)$ itself. On the last level ($i = r$), such a reduction via $J_r(q)$ is not required anymore. The complete, remaining null space $Y_{r-1}(q)$ can be

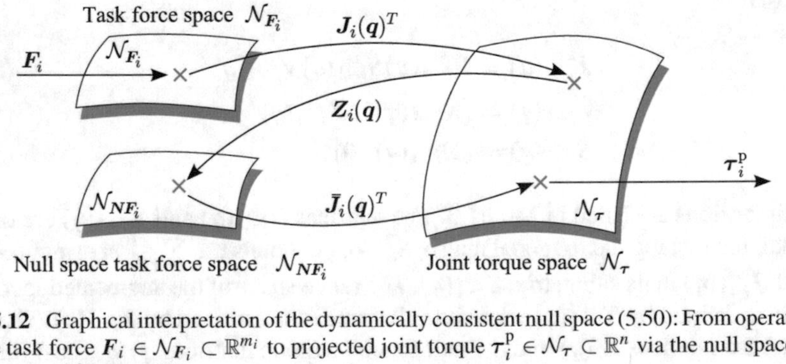

Fig. 5.12 Graphical interpretation of the dynamically consistent null space (5.50): From operational space task force $F_i \in \mathcal{N}_{F_i} \subset \mathbb{R}^{m_i}$ to projected joint torque $\tau_i^{\text{p}} \in \mathcal{N}_{\tau} \subset \mathbb{R}^n$ via the null space task force space $\mathcal{N}_{NF_i} \subset \mathbb{R}^{m_i}$

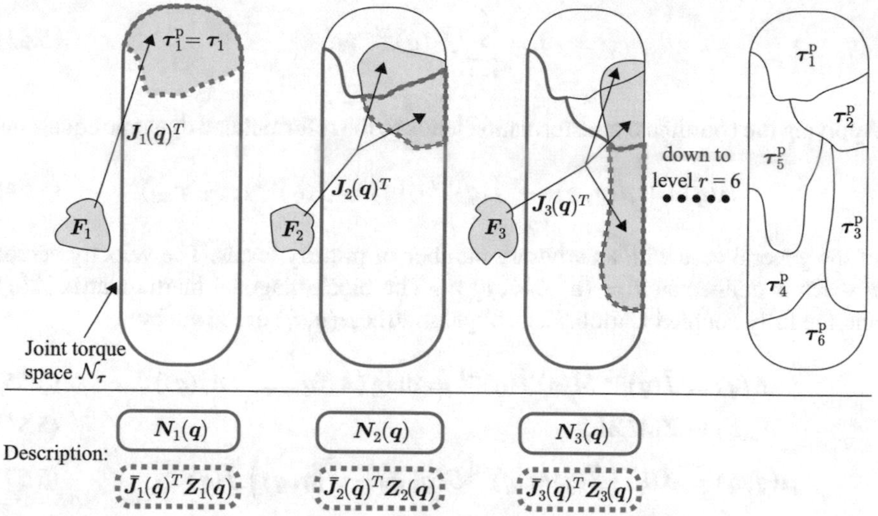

Fig. 5.13 Graphical interpretation of the null spaces for a hierarchy with $r = 6$ priority levels. Note that $N_1(q) = I$, since the main task is not restricted

directly applied as in the two-level case [OKN08]. Although the main task $i = 1$ is not constrained by any null spaces, $Z_1(q)$ is introduced to unify the notations.

Figure 5.13 depicts the null spaces obtained via (5.44) and (5.51). The figure illustrates the results of the mappings in Fig. 5.12. The main task force F_1 is mapped onto the joint torque space via $J_1(q)^T$ as shown in the first picture in Fig. 5.13. The task torque $J_2(q)^T F_2$ from level two (second picture), however, partially intersects the control torque of level one. This area is prohibited to comply with the order of priority, that is, the main task may not be disturbed. The null space projectors take that into account. One can see that the dynamically consistent null space projector $N_2(q)$ describes the complete null space of the main task. The subspace defined by the projector $\bar{J}_2(q)^T Z_2(q)$ leads to the same control input as $N_2(q)$, cf. (5.50), but it uses the additional information contained in $J_2(q)$ for a reduction to the subspace that can actually be used by the second-level task. The third picture describes the analogous relations for the third-level task, and the fourth picture illustrates the resulting, hierarchy-consistent distribution in the joint torque space. Both $N_i(q)$ and $\bar{J}_i(q)^T Z_i(q)$ lead to identical projected torques τ_i^p for $i = 1 \dots r$ (fourth picture), but $\bar{J}_i(q)$ and $Z_i(q)$ additionally define the corresponding task velocities v_i, which describe the decoupled dynamics and are required for the proof of stability.

Similar to (5.32), the inverse of $\bar{J}(q)$ in the multi-level case is given by

$$\bar{J}(q)^{-1} = \left(Z_1(q)^T, \; \dots, Z_r(q)^T\right) = \left(\bar{J}_1(q)^{M+}, \; \dots, \bar{J}_r(q)^{M+}\right). \qquad (5.52)$$

The proof of (5.52) can be found in Appendix C.2. The joint velocities can be expressed as the sum of the task velocities and the null space velocities:

$$\dot{\boldsymbol{q}} = \sum_{i=1}^{r} \bar{\boldsymbol{J}}_i(\boldsymbol{q})^{M+} \boldsymbol{v}_i. \tag{5.53}$$

Applying the coordinate transformation leads to the reformulated dynamic equations

$$\boldsymbol{\Lambda}(\boldsymbol{q})\dot{\boldsymbol{v}} + \boldsymbol{\mu}(\boldsymbol{q}, \dot{\boldsymbol{q}})\boldsymbol{v} + \bar{\boldsymbol{J}}(\boldsymbol{q})^{-T}\boldsymbol{g}(\boldsymbol{q}) = \bar{\boldsymbol{J}}(\boldsymbol{q})^{-T}(\boldsymbol{\tau} + \boldsymbol{\tau}_{\text{ext}}) \tag{5.54}$$

of the general case with an arbitrary number of priority levels. The velocity vector $\boldsymbol{v} \in \mathbb{R}^n$ is defined as $\boldsymbol{v} = (\boldsymbol{v}_1^T, \ldots, \boldsymbol{v}_r^T)^T$. The block-diagonal inertia matrix $\boldsymbol{\Lambda}(\boldsymbol{q})$ and the fully coupled Coriolis/centrifugal matrix $\boldsymbol{\mu}(\boldsymbol{q}, \dot{\boldsymbol{q}})$ are given by

$$\boldsymbol{\Lambda}(\boldsymbol{q}) = \bar{\boldsymbol{J}}(\boldsymbol{q})^{-T}\boldsymbol{M}(\boldsymbol{q})\bar{\boldsymbol{J}}(\boldsymbol{q})^{-1} = \text{diag}\left(\boldsymbol{\Lambda}_1(\boldsymbol{q}), \ldots, \boldsymbol{\Lambda}_r(\boldsymbol{q})\right), \tag{5.55}$$

$$\boldsymbol{\Lambda}_i(\boldsymbol{q}) = \boldsymbol{Z}_i \boldsymbol{M} \boldsymbol{Z}_i^T, \tag{5.56}$$

$$\boldsymbol{\mu}(\boldsymbol{q}, \dot{\boldsymbol{q}}) = \boldsymbol{\Lambda}(\boldsymbol{q})\left(\bar{\boldsymbol{J}}(\boldsymbol{q})\boldsymbol{M}(\boldsymbol{q})^{-1}\boldsymbol{C}(\boldsymbol{q}, \dot{\boldsymbol{q}}) - \dot{\bar{\boldsymbol{J}}}(\boldsymbol{q}, \dot{\boldsymbol{q}})\right)\bar{\boldsymbol{J}}(\boldsymbol{q})^{-1}. \tag{5.57}$$

The proof of the block-diagonality of $\boldsymbol{\Lambda}(\boldsymbol{q})$ can be found in Appendix C.3. Summarized, the dynamics representation in the particular form (5.54) is a generalization of (5.35) to a prioritized stack of tasks.

5.2.3 *Control Design*

The design aims at a compliance control on all hierarchy levels, where the measurement of the external forces is not required. However, keeping the natural inertia leads to a nonlinear closed-loop behavior. In the following, the control actions for the subtasks will be projected onto the respective null spaces according to the partially decoupled system dynamics (5.54). To prove stability, an additional feedback term is needed which brings the Coriolis/centrifugal matrix into a block-diagonal form. The control law is given by

$$\boldsymbol{\tau} = \sum_{i=1}^{r} \boldsymbol{\tau}_i^{\text{p}} + \boldsymbol{\tau}_{\mu} + \boldsymbol{g}(\boldsymbol{q}) \tag{5.58}$$

and contains the projected control actions $\boldsymbol{\tau}_1^{\text{p}}$ to $\boldsymbol{\tau}_r^{\text{p}}$ for the compliance control on all hierarchy levels,[9] a compensation of Coriolis and centrifugal couplings $\boldsymbol{\tau}_{\mu}$, and a gravity compensation term $\boldsymbol{g}(\boldsymbol{q})$. The components of the control action will be specified in the following. The signal flow chart in Fig. 5.14 depicts the closed loop.

---

[9]Note that the superscript "p" is also used for the main task with $\boldsymbol{\tau}_1^{\text{p}} = \boldsymbol{\tau}_1$ to unify the notations.

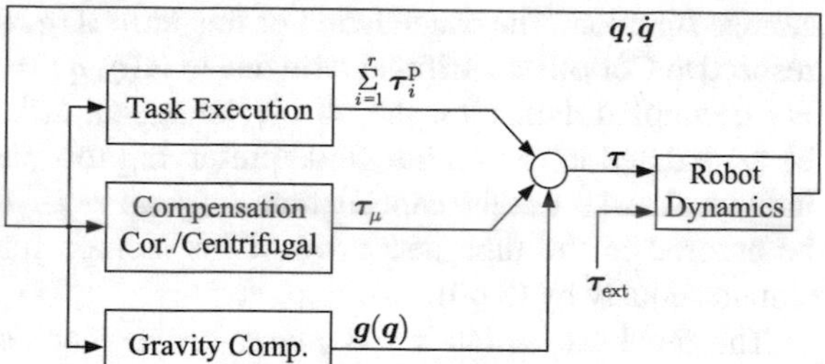

Fig. 5.14 Signal flow chart of the closed loop with controller (5.58)

5.2.3.1 Task Execution

The control input for task execution on all hierarchy levels ($1 \leq i \leq r$) is

$$\boldsymbol{\tau}_i^{\mathrm{p}} = -\bar{\boldsymbol{J}}_i(\boldsymbol{q})^T \boldsymbol{Z}_i(\boldsymbol{q}) \boldsymbol{J}_i(\boldsymbol{q})^T \underbrace{\left(\left(\frac{\partial V_i(\tilde{\boldsymbol{x}}_i)}{\partial \boldsymbol{x}_i} \right)^T + \boldsymbol{D}_i \dot{\boldsymbol{x}}_i \right)}_{\boldsymbol{F}_i}. \tag{5.59}$$

Note that for $i = 1$, (5.59) simplifies due to $\bar{\boldsymbol{J}}_1(\boldsymbol{q})^T \boldsymbol{Z}_1(\boldsymbol{q}) \boldsymbol{J}_1(\boldsymbol{q})^T = \boldsymbol{J}_1(\boldsymbol{q})^T$. The operational space force $\boldsymbol{F}_i \in \mathcal{N}_{F_i}$ is depicted in Fig. 5.12. The damping matrices $\boldsymbol{D}_i$ are positive definite and optionally configuration-dependent. The stability of this intuitive null space control law alone is not evident indeed. As a result of the null space projections via $\bar{\boldsymbol{J}}_i(\boldsymbol{q})^T \boldsymbol{Z}_i(\boldsymbol{q})$, the projected control torque does *not* represent a passive feedback action in general. This disadvantageous property is shared by *all* redundancy resolutions based on null space projections.

5.2.3.2 Power-Conserving Cancellation of Coriolis and Centrifugal Couplings

While the inertia matrix $\boldsymbol{\Lambda}(\boldsymbol{q})$ is already in block-diagonal form in (5.54), $\boldsymbol{\mu}(\boldsymbol{q}, \dot{\boldsymbol{q}})$ is still fully occupied. Therefore, Coriolis and centrifugal couplings have to be compensated for, i.e. the off-blockdiagonal entries in $\boldsymbol{\mu}(\boldsymbol{q}, \dot{\boldsymbol{q}})$. That is achieved by the feedback term

$$\boldsymbol{\tau}_{\mu} = \sum_{i=1}^{r} \left(\bar{\boldsymbol{J}}_i(\boldsymbol{q})^T \left(\sum_{j=1}^{i-1} \boldsymbol{\mu}_{i,j}(\boldsymbol{q}, \dot{\boldsymbol{q}}) \boldsymbol{v}_j + \sum_{j=i+1}^{r} \boldsymbol{\mu}_{i,j}(\boldsymbol{q}, \dot{\boldsymbol{q}}) \boldsymbol{v}_j \right) \right). \tag{5.60}$$

Herein $\boldsymbol{\mu}_{i,j}(\boldsymbol{q}, \dot{\boldsymbol{q}}) \in \mathbb{R}^{m_i \times m_j}$ denotes the submatrix of $\boldsymbol{\mu}(\boldsymbol{q}, \dot{\boldsymbol{q}})$ which is located in row block i and column block j. The reason for introducing (5.60) instead of cancelling the complete term $\boldsymbol{\mu}(\boldsymbol{q}, \dot{\boldsymbol{q}})$ is that any Lyapunov-based stability analysis incorporating the kinetic energy will inevitably produce $\dot{\boldsymbol{\Lambda}}(\boldsymbol{q}, \dot{\boldsymbol{q}})$ in the time derivative of the energy storage function. The cancellation of this term $\dot{\boldsymbol{\Lambda}}(\boldsymbol{q},\dot{\boldsymbol{q}})$ is usually conducted by the respective Coriolis/centrifugal term due to $\dot{\boldsymbol{\Lambda}}(\boldsymbol{q},\dot{\boldsymbol{q}}) = \boldsymbol{\mu}(\boldsymbol{q},\dot{\boldsymbol{q}}) + \boldsymbol{\mu}(\boldsymbol{q},\dot{\boldsymbol{q}})^T$, cf. (2.7). For decoupled dynamics, the off-blockdiagonal task couplings in $\boldsymbol{\mu}(\boldsymbol{q},\dot{\boldsymbol{q}})$ have to be annihilated in (5.54), but while preserving this passivity property related to the individual tasks, i.e. the cancellation $\dot{\boldsymbol{\Lambda}}_i(\boldsymbol{q},\dot{\boldsymbol{q}}) - \boldsymbol{\mu}_{i,i}(\boldsymbol{q},\dot{\boldsymbol{q}}) - \boldsymbol{\mu}_{i,i}(\boldsymbol{q},\dot{\boldsymbol{q}})^T = \boldsymbol{0}$ must be ensured in the time derivative of the storage function. Both can be achieved simultaneously by (5.60).

The feedback action $\boldsymbol{\tau}_\mu$ is *power-conserving* because the transmitted power $P_\mu = \boldsymbol{\tau}_\mu^T\dot{\boldsymbol{q}}$ is always zero. That is due to the skew symmetry $\boldsymbol{\mu}_{i,j}(\boldsymbol{q},\dot{\boldsymbol{q}}) = -\boldsymbol{\mu}_{j,i}(\boldsymbol{q},\dot{\boldsymbol{q}})^T$. This feature is very useful from a robustness point of view, since it is independent of parameter uncertainties in the model. While the dynamic effects related to $\boldsymbol{\mu}(\boldsymbol{q},\dot{\boldsymbol{q}})$ are of minor importance in practice for small to medium velocities, the feedback compensation (5.60) is required for the proof of stability in Sect. 5.2.4.

5.2.3.3 Closed-Loop Dynamics

In compliance control, the effect of the external forces and torques is of high interest. Since the augmented Jacobian matrix is non-singular, the term $\boldsymbol{\tau}_{\text{ext}}$ can be replaced by the components $\boldsymbol{F}_{i,\text{ext}} \in \mathbb{R}^{m_i}$ for $i = 1 \dots r$ related to the individual priority levels such that the following relation holds:

$$\boldsymbol{\tau}_{\text{ext}} = \bar{\boldsymbol{J}}(\boldsymbol{q})^T \begin{pmatrix} \boldsymbol{F}_{1,\text{ext}} \\ \vdots \\ \boldsymbol{F}_{r,\text{ext}} \end{pmatrix}. \tag{5.61}$$

Applying (5.58) finally yields the closed-loop dynamics

$$\boldsymbol{\Lambda}_i(\boldsymbol{q})\dot{\boldsymbol{v}}_i + \boldsymbol{\mu}_{i,i}(\boldsymbol{q},\dot{\boldsymbol{q}})\boldsymbol{v}_i + \boldsymbol{Z}_i(\boldsymbol{q})\boldsymbol{J}_i(\boldsymbol{q})^T \left(\left(\frac{\partial V_i(\tilde{\boldsymbol{x}}_i)}{\partial \boldsymbol{x}_i} \right)^T + \boldsymbol{D}_i\dot{\boldsymbol{x}}_i \right) = \boldsymbol{F}_{i,\text{ext}} \tag{5.62}$$

for $1 \le i \le r$ with $\boldsymbol{Z}_1(\boldsymbol{q})\boldsymbol{J}_1(\boldsymbol{q})^T = \boldsymbol{I}$. The change of the kinetic energy associated with each priority level only depends on the dynamics on this level. This is a direct consequence of (5.60). Note that the intuitive choice $S = \frac{1}{2}\boldsymbol{v}^T\boldsymbol{\Lambda}\boldsymbol{v} + \sum_{i=1}^r V_i(\tilde{\boldsymbol{x}}_i)$ does *not* represent a storage function of the closed-loop system due to consequences of the null space projections.

5.2.4 *Proof of Stability*

The dependencies on $\boldsymbol{q}$ are omitted in the notations for the sake of simplicity.

Proposition 5.1 *Consider the system (5.54) with the control law (5.58). The potential functions $V_i(\tilde{\boldsymbol{x}}_i)$ for $i = 1 \dots r$ are positive definite w.r.t. $\tilde{\boldsymbol{x}}_i$ and positive semi-definite w.r.t. $\boldsymbol{q}$. The damping matrices $\boldsymbol{D}_i$ for $i = 1 \dots r$ are positive definite. Then the closed-loop system is strictly output passive w.r.t. the input $\boldsymbol{F}_{1,\text{ext}}$ and the output $\dot{\boldsymbol{x}}_1$. Suppose also that the Jacobian matrices $\boldsymbol{J}_i$ for $i = 1 \dots r$ are of full row rank in the considered workspace, and $\bar{\boldsymbol{J}}$ is non-singular. The equilibrium $(\boldsymbol{q}^*, \boldsymbol{0})$, with $\boldsymbol{q}^*$ being the hierarchy-consistent, constrained local minimum of all V_i, is asymptotically stable for the case of free motion $\boldsymbol{\tau}_{\text{ext}} = \boldsymbol{0}$.*

The proof is based on semi-definite Lyapunov functions [IKO96] and *conditional stability* (see Appendix D.2). Additionally, the following theorem will be applied iteratively in the proof of stability.

Theorem 5.1 ([vdS00]) *Let the system*

$$\dot{z} = \boldsymbol{g}_1(z) + \boldsymbol{g}_2(z)\boldsymbol{u},$$
$$\boldsymbol{y} = \boldsymbol{h}(z)$$

with state $z \in \mathbb{R}^n$, input $\boldsymbol{u} \in \mathbb{R}^m$, and output $\boldsymbol{y} \in \mathbb{R}^m$ be strictly output passive for the output $\boldsymbol{y} = \boldsymbol{h}(z)$. Let further $\mathcal{A}$ be the largest positively invariant set contained in $\{z \in \mathbb{R}^n | \boldsymbol{h}(z) = \boldsymbol{0}\}$. If the equilibrium z^* is asymptotically stable conditionally to $\mathcal{A}$, then it is asymptotically stable for $\boldsymbol{u} = \boldsymbol{0}$.

In the proof of Proposition 5.1, several nested sets are used that represent the priority-consistent accomplishment of all involved tasks in the hierarchy. For the case of free motion ($\boldsymbol{F}_{1,\text{ext}} = \boldsymbol{0}$), the largest positively invariant set contained in $(\boldsymbol{q}, \boldsymbol{v}_1 = \boldsymbol{0}, \boldsymbol{v}_2, \dots, \boldsymbol{v}_r)$ is

$$\mathcal{A}_1 = \left\{(\boldsymbol{q}, \boldsymbol{v}_1, \dots, \boldsymbol{v}_r) | \boldsymbol{v}_1 = \boldsymbol{0}, \boldsymbol{f}_1(\boldsymbol{q}) = \boldsymbol{x}_{1,\text{des}}\right\}. \tag{5.63}$$

This set contains all system states where the first priority task has already been completed successfully. Now one can recursively define more sets to represent the additional accomplishment of the lower-priority tasks:

$$\mathcal{A}_i = \mathcal{A}_{i-1} \bigcap \left\{(\boldsymbol{q}, \boldsymbol{v}_1, \dots, \boldsymbol{v}_r) | \boldsymbol{v}_i = \boldsymbol{0}, \ \boldsymbol{Z}_i \boldsymbol{J}_i^T \left(\frac{\partial V_i(\tilde{\boldsymbol{x}}_i)}{\partial \boldsymbol{x}_i}\right)^T = \boldsymbol{0}\right\} \tag{5.64}$$

for $i = 2 \dots r - 1$. All sets are illustrated in Fig. 5.15 and describe the successive restriction to smaller subsets. The proof of asymptotic stability of the equilibrium $(\boldsymbol{q} = \boldsymbol{q}^*, \boldsymbol{v} = \boldsymbol{0})$ is based on the positive semi-definite storage functions

$$S_i = \frac{1}{2}\boldsymbol{v}_i^T \boldsymbol{\Lambda}_i \boldsymbol{v}_i + V_i(\tilde{\boldsymbol{x}}_i) \tag{5.65}$$

for all priority levels $i = 1 \dots r$. Using the property $\dot{\boldsymbol{\Lambda}}_i = \boldsymbol{\mu}_{i,i} + \boldsymbol{\mu}_{i,i}^T$ one can show that the time derivative of S_i along the solutions of the closed-loop system yields

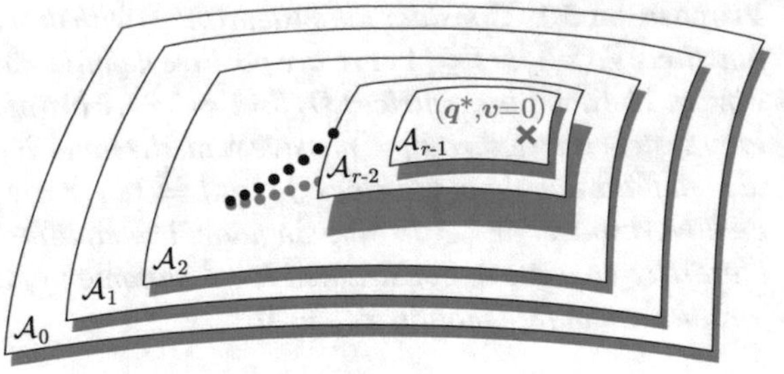

Fig. 5.15 Graphical interpretation of the sets used in the proof of stability. The largest set $\mathcal{A}_0$ represents the complete state space $(\boldsymbol{q}, \dot{\boldsymbol{q}})$. Each overlying set $\mathcal{A}_i$ is a subset of $\mathcal{A}_{i-1}$ for all $1 \leq i \leq r-1$

$$\dot{S}_i = \boldsymbol{v}_i^T \boldsymbol{F}_{i,\text{ext}} - \boldsymbol{v}_i^T \boldsymbol{Z}_i \boldsymbol{J}_i^T \boldsymbol{D}_i \boldsymbol{J}_i \dot{\boldsymbol{q}} - \boldsymbol{v}_i^T \boldsymbol{Z}_i \boldsymbol{J}_i^T \left(\frac{\partial V_i(\tilde{\boldsymbol{x}}_i)}{\partial \boldsymbol{x}_i}\right)^T + \left(\frac{\partial V_i(\tilde{\boldsymbol{x}}_i)}{\partial \boldsymbol{x}_i}\right) \boldsymbol{J}_i \dot{\boldsymbol{q}} \tag{5.66}$$

Within the set $\mathcal{A}_{i-1}$, the simplification

$$\boldsymbol{J}_i \dot{\boldsymbol{q}} = \boldsymbol{J}_i \boldsymbol{Z}_i^T \boldsymbol{v}_i \quad \forall (\boldsymbol{q}, \boldsymbol{v}) \in \mathcal{A}_{i-1} \tag{5.67}$$

holds thanks to (5.52) and (5.53) and the following two arguments: First, all contributions $\boldsymbol{v}_j$ from higher-priority tasks $(j < i)$ vanish due to the restriction to the set $\mathcal{A}_{i-1}$. Second, all contributions in $\dot{\boldsymbol{q}}$ that refer to lower-priority tasks $(k > i)$ vanish due to the orthogonality $\boldsymbol{J}_i \boldsymbol{Z}_k^T = \boldsymbol{0}$. Hence, (5.66) yields

$$\dot{S}_i = \boldsymbol{v}_i^T \boldsymbol{F}_{i,\text{ext}} \underbrace{- \boldsymbol{v}_i^T \boldsymbol{Z}_i \boldsymbol{J}_i^T \boldsymbol{D}_i \boldsymbol{J}_i \boldsymbol{Z}_i^T \boldsymbol{v}_i}_{\leq 0} \quad \forall (\boldsymbol{q}, \boldsymbol{v}) \in \mathcal{A}_{i-1}. \tag{5.68}$$

Proof The line of argumentation starts in the set $\mathcal{A}_{r-1}$, i.e. all tasks but the lowest-priority task are assumed to be accomplished already as illustrated in Fig. 5.15. Now consider (5.65) and (5.68) for $i = r$. From (5.68) one can conclude conditional stability w.r.t. the set $\mathcal{A}_{r-1}$ for the case of free motion $\boldsymbol{F}_{r,\text{ext}} = \boldsymbol{0}$. According to LaSalle's invariance principle, the state converges to the largest positively invariant set contained in $\mathcal{A}_{r-1}$ where $\boldsymbol{v}_r = \boldsymbol{0}$. By investigating (5.62) one can see that this set requires

$$\boldsymbol{Z}_i \boldsymbol{J}_i^T \left(\frac{\partial V_i(\tilde{\boldsymbol{x}}_i)}{\partial \boldsymbol{x}_i}\right)^T = \boldsymbol{0} \tag{5.69}$$

for all levels $i = 1 \ldots r$. If the tasks partially conflict with each other, a constrained local minimum $\boldsymbol{q}^*$ is reached which complies with the order of priority.[10] Although $\partial V_i(\tilde{\boldsymbol{x}}_i)/\partial \boldsymbol{x}_i \neq \boldsymbol{0}$ on at least one level i, (5.69) is fulfilled due to the annihilation via $\boldsymbol{Z}_i$. The interpretation of (5.69) is that it constitutes the condition for static consistency as defined in Definition 4.1. One can conclude asymptotic stability of $(\boldsymbol{q}^*, \boldsymbol{0})$ conditionally to the set $\mathcal{A}_{r-1}$.

[10] If the tasks do not even statically conflict with each other at all, then $\boldsymbol{q}^*$ is actually the global minimum of all V_i. In that case, (5.69) is fulfilled because $\partial V_i(\tilde{\boldsymbol{x}}_i)/\partial \boldsymbol{x}_i = \boldsymbol{0} \; \forall i$.

Now Theorem 5.1 can be applied within the set $\mathcal{A}_{r-2}$ for $\mathcal{A} = \mathcal{A}_{r-1}$, $\boldsymbol{u} = \boldsymbol{F}_{r-1,\text{ext}}$, and $\boldsymbol{y} = \boldsymbol{v}_{r-1}$. Strict output passivity is given by (5.68) for $i = r - 1$, and asymptotic stability conditionally to $\mathcal{A}_{r-1}$ has already been shown. That allows to conclude asymptotic stability of $(\boldsymbol{q}^*, \boldsymbol{0})$ for $\boldsymbol{F}_{r-1,\text{ext}} = \boldsymbol{0}$. Note again that this conclusion is only valid within the set $\mathcal{A}_{r-2}$, thus is it also of *conditional stability* nature only.

As of now, Theorem 5.1 can be iteratively applied, beginning within the set $\mathcal{A}_{r-3}$ for $\mathcal{A} = \mathcal{A}_{r-2}$, $\boldsymbol{u} = \boldsymbol{F}_{r-2,\text{ext}}$, and $\boldsymbol{y} = \boldsymbol{v}_{r-2}$, i.e. $i = r - 2$ in (5.68). In each iteration step, one can conclude strict output passivity with (5.68). Together with the conditional asymptotic stability obtained in the previous iteration step, Theorem 5.1 allows to conclude asymptotic stability of $(\boldsymbol{q}^*, \boldsymbol{0})$ conditionally to $\mathcal{A}_{r-3}$ for the case of free motion $\boldsymbol{F}_{r-2,\text{ext}} = \boldsymbol{0}$. This iterative application of Theorem 5.1 can be performed straightforwardly up to the main task level. There, the conditional stability w.r.t. the set $\mathcal{A}_1$, together with the passivity property of the main task, proves asymptotic stability of $(\boldsymbol{q}^*, \boldsymbol{0})$ for the case of free motion. The strict output passivity claimed in Proposition 5.1 has been used in this last step of the proof and can be verified by evaluating (5.68) for $i = 1$. □

5.2.5 *Simulations and Experiments*

Three simulations on a planar $n = 4$ DOF system and two experiments on Rollin' Justin are conducted to validate the decoupling, the stability properties, the performance in case of model uncertainties, and the practical influence of $\boldsymbol{\tau}_{\mu}$.

A schematic representation of the simulated dynamic system is given in Fig. 4.1. The task hierarchy in Table 5.4 is applied, and the controller gains are specified in Table 5.5. The approach is evaluated for step responses on all hierarchy levels.

5.2.5.1 Simulation #1: Decoupling

In simulation #1, the damping is set very high and the task execution as well as the decoupling quality are analyzed. The errors in the operational space are provided in Fig. 5.16. They all converge to zero within less than 0.5 s. The corresponding control torques on the three levels are depicted in Fig. 5.17 (left). After solving the dynamics

Table 5.4 Initial conditions and desired values for the simulations

Priority	Description	Task var.	Initial val.	Goal val.
Level 1	Translational compliance at TCP	x of TCP	0.84 m	0.90 m
		y of TCP	0.96 m	0.80 m
Level 2	Rotational compliance at TCP	$\sum_{i=1}^{4} q_i$	−1.37 rad	−1.57 rad
Level 3	Joint compliance in first joint	q_1	0.40 rad	0.35 rad

Table 5.5 Controller parameterization for the simulations

Priority	Gain	Unit	Simulation #1	Simulation #2	Simulation #3
Level 1	$\boldsymbol{K}_1$	$\frac{\mathrm{N}}{\mathrm{m}}$	diag(1000, 1000)	diag(500, 500)	diag(500, 500)
	$\boldsymbol{D}_1$	$\frac{\mathrm{Ns}}{\mathrm{m}}$	diag(40, 40)	diag(30, 30)	diag(30, 30)
Level 2	$\boldsymbol{K}_2$	$\frac{\mathrm{Nm}}{\mathrm{rad}}$	800	600	600
	$\boldsymbol{D}_2$	$\frac{\mathrm{Nm\cdot s}}{\mathrm{rad}}$	5	1.5	1.5
Level 3	$\boldsymbol{K}_3$	$\frac{\mathrm{Nm}}{\mathrm{rad}}$	2400	600	600
	$\boldsymbol{D}_3$	$\frac{\mathrm{Nm\cdot s}}{\mathrm{rad}}$	15	1.5	1.5

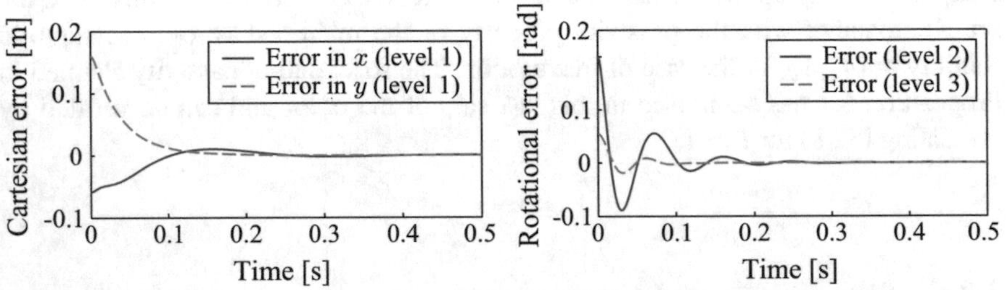

Fig. 5.16 Simulation #1: The errors in the operational spaces of the hierarchy levels converge to zero. Initial and goal values are summarized in Table 5.4

(5.62) for the accelerations, one can evaluate the undesired effects induced by active control on the other priority levels as plotted in Fig. 5.17 (right). Dynamic consistency is achieved due to the decoupling, a strict hierarchy realized.

5.2.5.2 Simulation #2: Stability Properties

In simulation #2, the damping is set very low in order to demonstrate the stability properties without eliminating undesired (velocity-dependent) effects by energy dissipation through damping injection. Figure 5.18 (left) shows the errors on all three levels. The main task is undisturbed and converges. At about $t = 0.4\,\mathrm{s}$, the primary task error is almost zero (top) and it is not affected by the remaining null space motions (bottom). The latter require a longer time to get into a steady state at zero. The corresponding energies are plotted in Fig. 5.18 (right). In accordance to the primary task error, the total energy related to the main task tends to zero before convergence on the lower-priority levels can be observed. The center chart and the bottom chart illustrate the energy contributions on the two null space levels. The total energies on level two and level three do not monotonically decrease. That complies with the stability properties due to conditional stability. In consideration of the fact that the total energy of the highest-priority task requires $0.2\,\mathrm{s}$ to reach "almost" zero, the behavior on level two is as expected. After $t = 0.2\,\mathrm{s}$, the total energy on level two monotonically decreases. The same applies to the third level w.r.t. the second

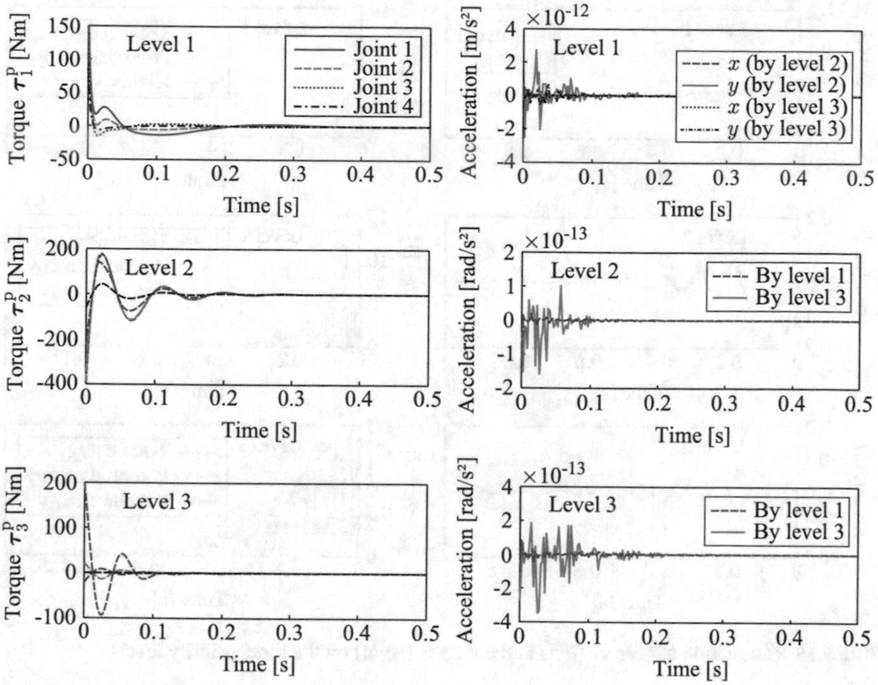

Fig. 5.17 Simulation #1: Contributions of the control input (*left*) and disturbing accelerations (*right*) induced by them

level after about $t = 0.4\,\mathrm{s}$. The stability analysis implies a monotonically decreasing energy as soon as the second level is converged.

5.2.5.3 Simulation #3: Model Uncertainties

Decoupled dynamics as formulated in (5.62) can only be preserved if the null space projectors $\bar{\boldsymbol{J}}_i(\boldsymbol{q})^T \boldsymbol{Z}_i(\boldsymbol{q})$ are dynamically consistent according to Definition 4.2, i.e. the model-based inertia matrix used in the control design perfectly matches the one of the actual dynamics. Simulation #3 addresses the robustness of the proposed controller under modeling uncertainties. The parameters are taken from Tables 5.4 and 5.5 again, but the assumed inertia in the controller differs in a way such that the first two link masses have been reduced by 10 %, and the last two link masses have been increased by 10 %. Additionally, each point mass has been shifted by 5 cm (closer to the TCP). Due to the error in the inertia matrix, a full decoupling cannot be achieved, and the control torques from each level lead to acceleration errors on the remaining hierarchy levels. Figure 5.19 demonstrates the influence of the level three task on the other tasks. The acceleration errors in the operational spaces of the respective task levels are plotted. Note that there also exists an effect from the level two control input on level one and level three as well as an effect from the

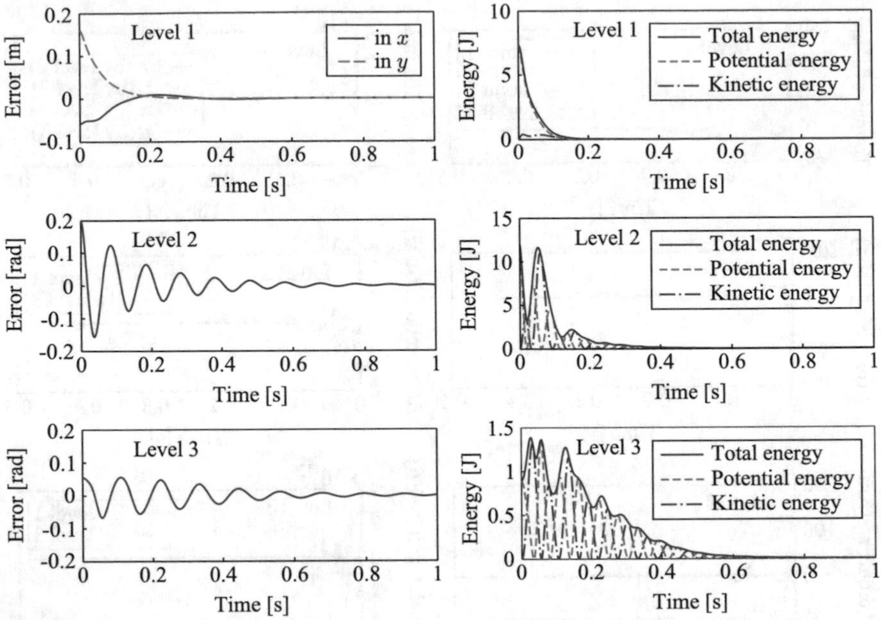

Fig. 5.18 Simulation #2: errors (*left*) and energies (*right*) on the three priority levels

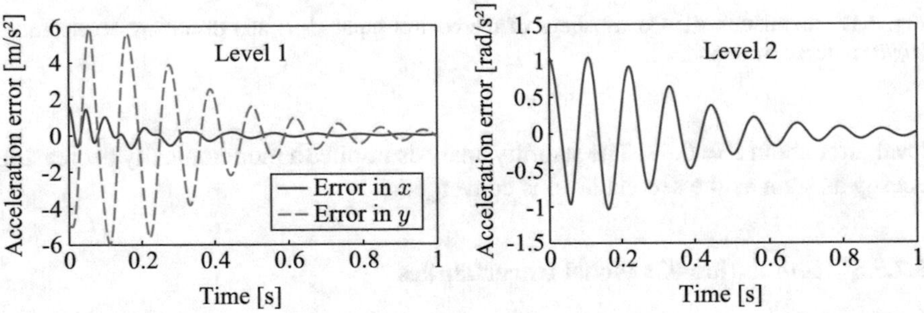

Fig. 5.19 Simulation #3: Instantaneous acceleration errors in the operational space induced by the level three torques. These undesired couplings occur because the inertia matrix used in the controller does not match the real one

main task control input on the two lower-priority levels. However, the closed-loop system is still robust with respect to disturbances in the model. To emphasize that statement, the control errors for simulation #3 are depicted in Fig. 5.20. Despite the large interferences, the transient behavior coincides well with the case study with undisturbed inertia matrix before. It goes without saying that this simulation cannot assess the robustness of the approach in general. But the behavior indicates a certain degree of robustness. The following experiments on Rollin' Justin confirm that conclusion.

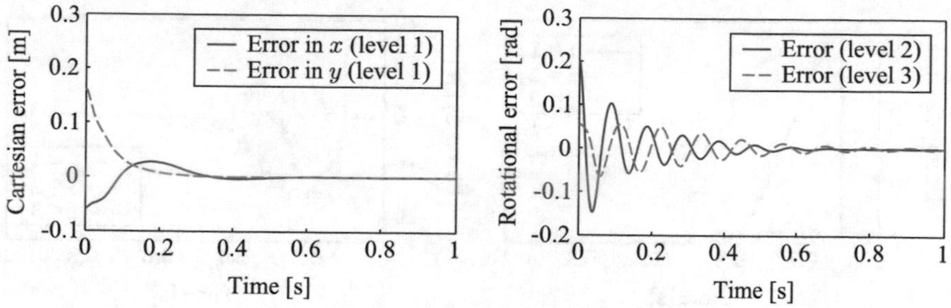

Fig. 5.20 Simulation #3: Errors on the three priority levels. Although the inertia matrix used in the controller significantly differs from the actual one, the transient behavior resembles the undisturbed case in Fig. 5.18

5.2.5.4 Experiment #1: Performance on Rollin' Justin

The controller is implemented on the arm of Rollin' Justin with seven actuated DOF and a task hierarchy with three priority levels. The main task is a translational Cartesian compliance at the TCP. On level two, the orientation of the TCP is assigned with a specified impedance behavior. Lowest priority is given to a joint compliance of the whole arm with fixed equilibrium configuration. The parameterization of the controller is provided in Table 5.6. Note that the damping factors $\boldsymbol{\xi}_1$ and $\boldsymbol{\xi}_2$ determine the positive definite damping matrices $\boldsymbol{D}_1(\boldsymbol{q})$ and $\boldsymbol{D}_2(\boldsymbol{q})$. The damping matrices are computed via the *Double Diagonalization* approach by assigning a desired mass-spring-damper relation taking into account the reflected inertias and the stiffness matrices [ASOFH03]. The step responses are analyzed in the following. Both the desired equilibria of the first and the second level task are switched at $t = 0.1$ s. The desired equilibrium of the third level task is kept constant. Hence, the final steady state is not compatible with all tasks simultaneously. There exists a joint configuration $\boldsymbol{q}^*$ that fulfills $\boldsymbol{x}_{i,\text{des}} = \boldsymbol{f}_i(\boldsymbol{q}^*)$ for $i = 1, 2$, but there is no configuration $\boldsymbol{q}^*$ that complies with $\boldsymbol{x}_{i,\text{des}} = \boldsymbol{f}_i(\boldsymbol{q}^*)$ for $i = 1, 2, 3$. The configuration converges to a constrained local minimum of the level three task. Intuitively, the controller approaches an overall

Table 5.6 Controller parameterization for the experiments

Priority	Gain	Value	Unit
Level 1	$\boldsymbol{K}_1$	diag(1200, 1200, 1200)	$\frac{\text{N}}{\text{m}}$
	$\boldsymbol{\xi}_1$	(0.9, 0.9, 0.9)	–
Level 2	$\boldsymbol{K}_2$	diag(30, 30, 30)	$\frac{\text{Nm}}{\text{rad}}$
	$\boldsymbol{\xi}_2$	(0.9, 0.9, 0.9)	–
Level 3	$\boldsymbol{K}_3$	diag(10, 10, 10, 10, 10, 10, 10)	$\frac{\text{Nm}}{\text{rad}}$
	$\boldsymbol{D}_3$	diag(3, 3, 3, 3, 3, 3, 3)	$\frac{\text{Nm·s}}{\text{rad}}$

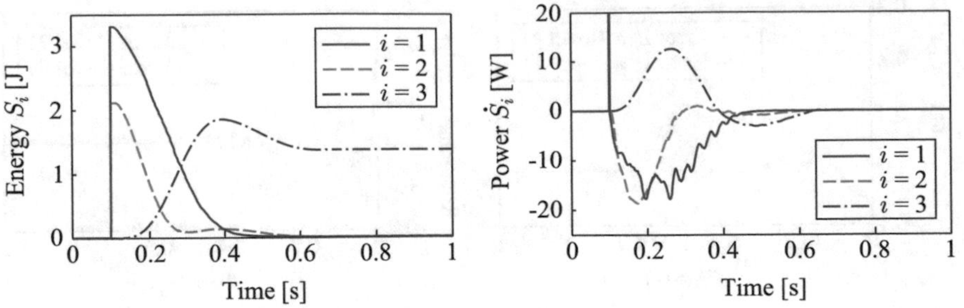

Fig. 5.21 Experiment #1: storage functions and their time derivatives on the three priority levels

equilibrium that complies with the order of priority and completes the tasks on level one and two, and it will execute the task on level three in the best possible way.

The storage functions (5.65) for all levels are plotted in Fig. 5.21 (left). The main task can be completed undisturbed. In Fig. 5.21 (right) one can see that the time derivative of S_1 is always lower than or equal to zero after the step. The energy on level two asymptotically converges to zero but it temporarily increases during $0.3\,\text{s} < t < 0.4\,\text{s}$. This behavior is consistent with the stability properties established before. The subordinate task on the third level cannot be accomplished at all. However, the remaining one-dimensional null space of the Cartesian TCP impedances allows to reach a constrained local minimum, which can be observed in the steady state behavior for $t \to \infty$. The operational space errors are depicted in Fig. 5.22. For the sake of simplicity, the three translational errors are summarized in the total Euclidean error. Likewise, the error on level two represents the absolute angle between actual and desired TCP frame. The error on level three describes the unitless Euclidean norm of the vector of the joint errors. The steady-state errors on the first two levels result from uncompensated friction and model uncertainties. Bear in mind that the fundamental structure of the compliance controller basically constitutes a PD controller, cf. Sect. 2.2.1.

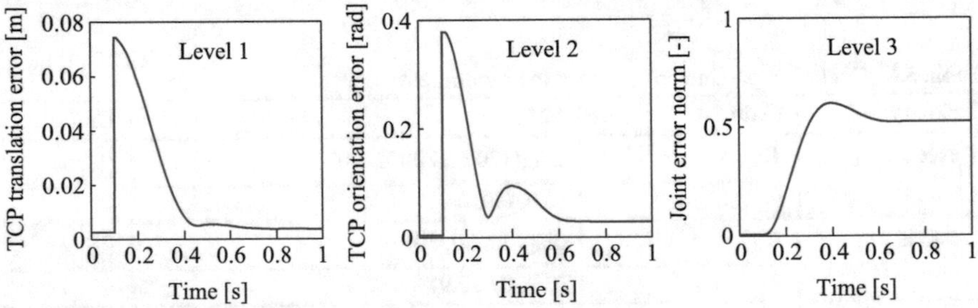

Fig. 5.22 Experiment #1: errors on the three priority levels

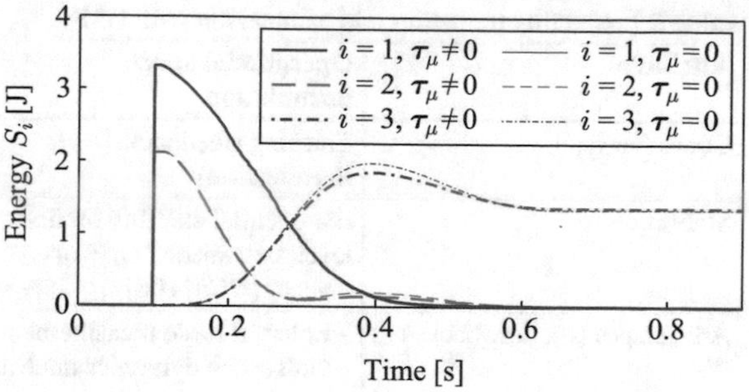

Fig. 5.23 Experiment #2: Direct comparison between active ($\boldsymbol{\tau}_\mu \neq \mathbf{0}$) and inactive ($\boldsymbol{\tau}_\mu = \mathbf{0}$) cancellation of Coriolis and centrifugal couplings. The influence of the compensation is very small in the considered setup

5.2.5.5 Experiment #2: Practical Influence of the Coriolis/Centrifugal Couplings

The previous experiment is repeated without the compensation of the Coriolis/centrifugal couplings in order to investigate the practical influence of (5.60). Figure 5.23 depicts the storage functions for active ($\boldsymbol{\tau}_\mu \neq \mathbf{0}$, identical to Fig. 5.21) and inactive ($\boldsymbol{\tau}_\mu = \mathbf{0}$) cancellation. The behaviors are very similar. Numerous runs revealed that the deviations can be hardly traced back to the term $\boldsymbol{\tau}_\mu$. To support that statement, the control torques can be consulted. The torque on level one, for example, attains a maximum value of about 28.4 Nm. In contrast, the maximum torque in $\boldsymbol{\tau}_\mu$ is less than 1.8 Nm. Although essential for the proof of stability, the practical influence of the compensation is very limited.

5.2.6 *Discussion*

The iterative application of Theorem 5.1 in the proof of stability can be interpreted as a sequential convergence of the different tasks according to their priority levels. Against intuition, Theorem 5.1 does *not* require finite time for reaching $\mathcal{A}$ in order to conclude asymptotic stability. Therefore, concluding asymptotic stability by iteratively applying it does not involve any successive finite time convergence.

In addition to asymptotic stability, the line of argumentation allows a hierarchical passivity statement:

Proposition 5.2 *Consider the closed-loop system (5.62) with positive definite potential functions $V_j(\tilde{\boldsymbol{x}}_j)$ $\forall j = 1 \ldots r$. Under the assumption that all external forces $\boldsymbol{F}_{k,\text{ext}}$ $\forall k < i$ are zero, then the system is strictly output passive conditionally to $\mathcal{A}_{i-1}$ w.r.t. the input $\boldsymbol{F}_{i,\text{ext}}$ and the output $\boldsymbol{v}_i$.*

The definition of strict output passivity conditionally to a set is given in Appendix D.1. One can even state that the higher-priority tasks do not necessarily need to reach the desired, hierarchy-consistent equilibrium positions; any steady-state configuration (reached for a given static disturbance) is sufficient to conclude conditional strict output passivity.

Table 5.7 Stability properties and comparison with OSF

Attribute	Operational space formulation	New approach
Controller type	Tracking (feedback linearization)	Compliance
Stability	Exponential stability on first level, null space behavior unclear [NCM+08]	Asymptotic stability, successive convergence, passivity-based
Advantages (+), drawbacks (–)	– External force measurements – Null space dynamics unclear + Tracking performance	+ No external force measurements + Stability includes null space – No tracking
Limitations in application	External forces in null space difficult to acquire	Proof of stability only valid for regulation case, not for tracking
Qualification	Trajectory tracking	Interaction tasks

The matrix $\bar{J}$ in (5.41) is assumed to be non-singular. Although many techniques exist to deal with singularities [CK95, MRC09, DWASH12a], the integration of these concepts in a closed-loop stability analysis has not been conducted yet and is topic of current research efforts. Nevertheless, a specific case of singularity can be dealt with here: If two or more tasks are always in conflict, e.g. their task directions are identical but they are placed on different levels in the hierarchy, that issue can be solved by eliminating these task directions from the lower-priority levels via appropriate preprocessing.

A detailed comparison between classical force-based operational space controllers (based on the OSF [Kha87]) and the approach presented here is drawn in Table 5.7. Concerning stability, the classical approaches provide exponential convergence on the first level. However, as stated by Nakanishi et al. [NCM+08], a formal stability statement for all lower levels is not known so far. The authors summarize "*the exact behavior of the null space dynamics cannot be determined easily (...) This difficulty of understanding the null space stability properties is, however, a problem that is shared by all operational space controllers. So far, only empirical evaluations can help to assess the null space robustness*". Here, asymptotic stability is shown by means of passivity theory and a conditional stability theorem. Moreover, the approach does not require feedback of the external forces and torques. The major restriction of the feedback of external forces and torques in the classical approaches has already been analyzed [PAW10, OKN08], and it poses problems in terms of robustness and availability of measurements. In conclusion, one can say that the classical feedback linearization is well suited for trajectory execution without contacts, i.e. external forces and torques do not have to be taken into account. The approach proposed here is better suited for contacts and interaction tasks where physical compliance is needed. Stability statements can be given for the overall system and the passivity features qualify for interactions in dynamic and unknown environments.

Another class of hierarchy-based approaches are the concepts by Mansard et al. [Man12, MKK09]. Compared to the approach presented here, they formulate an explicit QP (quadratic programming) problem with kinematic tasks to be solved numerically. The major advantages are that inequality constraints can be directly incorporated in the optimization problem and singularities can also be dealt with. But one has to compute the complete inverse dynamics, external interaction forces and torques are not considered, and a stability analysis has not been conducted yet.

5.3 Summary

The stability of the whole-body controller for wheeled humanoid robots was investigated in this chapter. To this end, two self-contained stability analyses had to be performed, which were then combined to one unified framework.

The first stability analysis referred to the case of a torque-controlled upper body which is mounted on a kinematically controlled mobile base. Since the mobile platform is only able to realize motions instead of forces and torques, the goal of Sect. 5.1 was to provide the means for an overall force-torque-based impedance controller. An admittance interface to the mobile base was utilized, and by modification of the dynamic couplings between the subsystems, the overall dynamics could be altered such that asymptotic stability of the desired equilibrium could be shown. The results of Sect. 5.1 can be interpreted as a fully torque-controlled wheeled robot with proven stability and undefined null space.

The second stability analysis in Sect. 5.2 referred to a generic torque-controlled robot with multiple simultaneous objectives. In this respect, it addressed the priority-based redundancy resolution from Chap. 4. The main result of Sect. 5.2 is the proof of stability for a task hierarchy with an arbitrary number of priority levels. A new representation of the equations of motion considering the hierarchical dynamics was presented. This formulation in combination with the theory of conditional stability made it possible to conclude asymptotic stability of the equilibrium and passivity properties.

The outcome of Sect. 5.1, that is a wheeled humanoid robot with overall force-torque interface, proven stability, but undefined null space behavior, can be used in combination with the generic multi-objective control in Sect. 5.2 to define this null space. Then the stability analysis in Sect. 5.2 applies to the complete *mobile* humanoid robot.

Chapter 6
Whole-Body Coordination

As a result of intensive research over the last decades, several robotic systems are approaching a level of maturity that allows robust task execution and safe interaction with humans and the environment. Besides humanoid robots such as ASIMO [SWA+02], Robonaut 2 [DMA+11], or HRP-4 [KKM+11], a variety of wheeled systems has been developed, e.g. Rollin' Justin [BWS+09], ARMAR-III [ARS+06], TWENDY-ONE [IS09], PR2 [BRJ+11]. Regardless of the specific structure of the system, the requirement of handling several objectives simultaneously is a common property for robotic applications in these dynamic and often unstructured environments. The features range from precise task execution, collision avoidance, and the compliance with physical constraints, to objectives such as maintaining the manipulability or the realization of desired postures.

Based on Khatib's *Operational Space Formulation* [Kha87], many different methods have been developed for planning and reactive control of such systems [BKV02, KSPW04, SK05, SGJG10, NKS+10, BHG10, HOD10, DWAS11, DWASH12b, SRK+13, LVYK13]. Due to the large number of approaches, details of the individual concepts are not presented here. Nevertheless, some basic tendencies can be pointed out: The majority of the approaches rest upon the design of artificial repulsive/attractive potential fields [Kha86]. Furthermore, the *Elastic Strips* framework by Brock and Khatib is frequently implemented in whole-body controllers in order to execute previously planned motions in a dynamic environment [BK02]S. The authors reactively adapt to changes in the environment, e.g. when an obstacle is approaching the manipulator. Here, the whole-body control framework [DWAS11, DWASH12b], which is based on the results of the previous chapters, is implemented and validated on Rollin' Justin. The major differences to the state of the art are: The considered robotic system is torque-controlled, whereas most state-of-the-art approaches still apply pure kinematic controllers. The experimental validation is the focus in Chap. 6. Although several other approaches seem promising in terms of real applications, by now they have been tested in simulations only.

A. Dietrich, *Whole-Body Impedance Control of Wheeled Humanoid Robots*,
Springer Tracts in Advanced Robotics 116, DOI 10.1007/978-3-319-40557-5_6

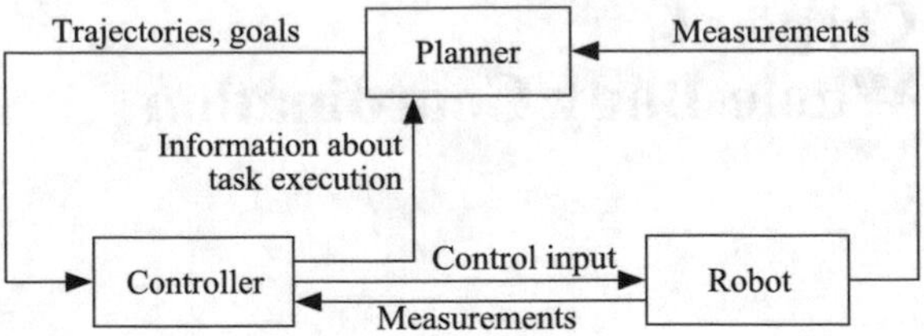

Fig. 6.1 Generic control flow for whole-body coordination

Section 6.1 treats the installation of a priority list that contains aspects of safety, physical constraints, task execution, and optimization criteria. In Sect. 6.2, the generic case of whole-body coordination (Fig. 6.1) is specified to yield a controller for the evaluation on Rollin' Justin. The experiments do not incorporate the feedback from controller to planner yet. That topic is deferred to Chap. 7.

6.1 Order of Tasks in the Hierarchy

In Chap. 4, a redundancy resolution has been proposed which establishes a priority-based hierarchy among a variety of tasks such as the ones in Chap. 3. But which criteria are relevant for the choice and prioritization of the involved tasks?

No matter how many subtasks are defined, safety aspects are of top priority. That comprises the safety of humans in the workspace of the robot as well as the environment itself. If that requirement is met, hard physical constraints should be addressed and the execution of the tasks can be considered. If sufficient structural redundancy is left, further subtasks can be carried out, e.g. desired postures or the optimization of the energy efficiency. In summary, one can establish the overall hierarchy:

1. Safety
2. Physical constraints
3. Task execution
4. Optimization criteria

Considering such a guideline for the priority list is the intuitive basis of many whole-body control approaches such as [SK05]. Figure 6.2 illustrates the different domains. In the following, they will be detailed.

Safety

As Isaac Asimov stated in his 1st law in 1942 [Asi42]: "*A robot may not injure a human being or, through inaction, allow a human being to come to harm*". Besides applying planning strategies to prevent dangerous situations in advance, the robot must be capable of feeling contact forces so as to react properly if a situation with physical human-robot interaction occurs [HASH09].

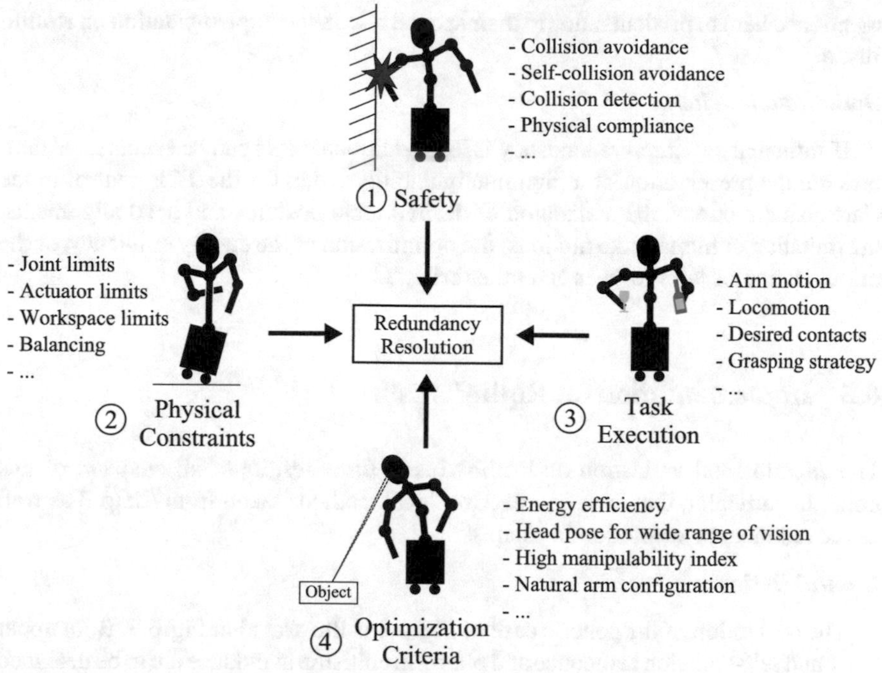

Fig. 6.2 Order of tasks in the hierarchy. Four basic domains with different priorities are distinguished here: safety, physical constraints, task execution, and optimization criteria

Among others, the following tasks can be enumerated: Collision avoidance with the environment, self-collision avoidance (Sect. 3.1), collision detection and softening, and triggered emergency stops.

Physical Constraints

This aspect refers to physical limitations and restrictions of the robotic system. That includes hardware-related issues such as the avoidance of mechanical end stops of joints (Sect. 3.4.3), the compliance with actuator limits (saturations, maximum joint torques, maximum joint velocities (Sect. 3.2)), kinematic constraints (Sect. 3.3), or balancing of legged robots on bumpy or flexible ground. It should be remarked that the distinction between safety aspects and physical constraints is ambiguous in some cases. For example, keeping the balance in a legged humanoid robot is a physical constraint, but it may lead to a safety risk if not executed properly.

Task Execution

Among others, this category includes the TCP behavior, e.g. physical contacts or the realization of TCP trajectories. Other important aspects are the locomotion or grasping of objects and dexterous manipulation using the available hardware such as hands or grippers. While task execution may comprise a large number of concepts, the ones based on impedances (Sect. 2.2.1) and admittances (Sect. 2.2.2) are to be mentioned here in particular due to their special role in the implementation on Rollin' Justin.

Optimization Criteria

If sufficient structural redundancy is left, additional tasks can be executed. Examples are the preservation of a high manipulability index for the TCP control in the Cartesian directions, the realization of desired torso postures and head alignments, the imitation of human-like motions, the optimization of the energy efficiency, or the minimization of joint torques and other criteria.

6.2 Implementation on Rollin' Justin

The experimental validation on Rollin' Justin fuses results of all chapters of this book. In particular, that includes reactive, local methods taken from Chap. 3 as well as the redundancy resolution in Chap. 4.

Control Design

The adaptation of the generic case of Fig. 6.1 is illustrated in Fig. 6.3. Both upper body and self-collision avoidance and platform collision avoidance are to be designed twice. In the first case, tight and very strong repulsion fields are given top priority, together with the task execution. The second case, which includes weaker and largely

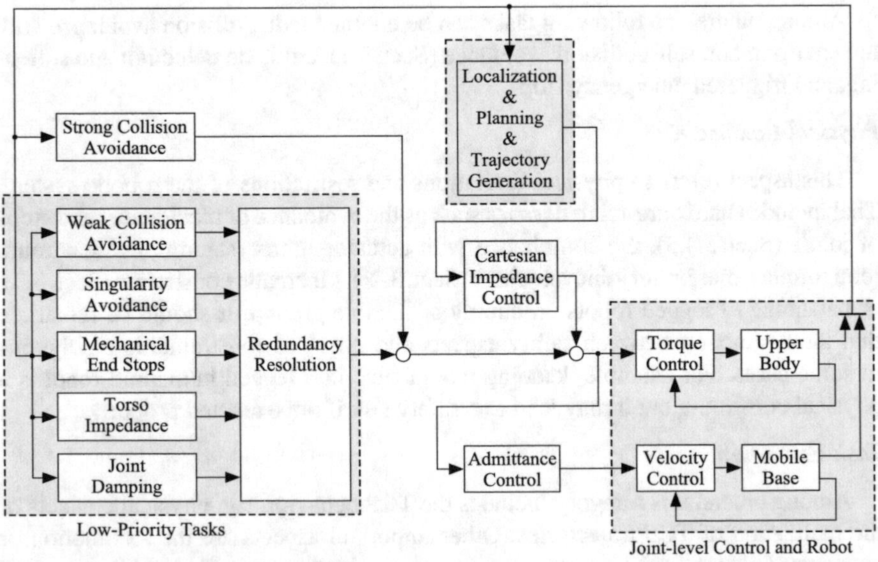

Fig. 6.3 Whole-body control implementation on Rollin' Justin. The lower-priority level is defined by the null space of the Cartesian impedance of the TCP

extended fields, is used on a lower-priority level. The kinematic redundancy of the Cartesian impedance is exploited to keep the robot in a "good" configuration. The task execution is not disturbed by the strong collision avoidances except for critical configurations. Then, the strong collision avoidance gets activated, it outplays the Cartesian impedance on the same hierarchy level and ensures safety. The arm singularity avoidance, the platform singularity avoidance, the avoidance of mechanical end stops, the torso impedance, and the joint damping, have lower priority.

In this redundancy resolution, some physical constraints are given a lower priority than the task execution. This is due to the fact that the robot has a very large number of actuated DOF. Therefore, sufficient kinematic redundancy is left for these lower-priority tasks without disturbing the main task. As a result of the design of the subtasks and the hierarchy, local minima can be largely avoided [DWASH12b]. Nevertheless, one has to keep in mind that this redundancy resolution implemented on Rollin' Justin is not generic but it is optimized for the particular characteristics of the robot.

Step Response

In experiment #1, the step response of the right TCP in the case of a forward motion ($\Delta x = 0.2\,\text{m}$) is evaluated. All subtasks are activated and the translational stiffnesses of the Cartesian impedance are set to $k_{\text{tra}} = 500\,\text{N/m}$. As it can be seen in Fig. 6.4, the actual settling time is less than 0.5 s. Besides, an overshooting is mentionable, which can be traced back to the delayed behavior of the platform due to the admittance coupling, see Fig. 6.4 (bottom), and a damping ratio of $\xi = 0.7$ in the impedance. As the impedance is basically a PD-controller and does not possess an integrating component, a steady-state error may remain (upper plots). The excitation in x-direction also affects the other two translational directions marginally. The steady-state errors can be reduced by applying a higher translational stiffness.

Subtask Contributions

In experiment #2, a continuous trajectory of the TCP is applied according to the upper chart in Fig. 6.5. The initial configuration of the robot is depicted in Fig. 6.6a. The right TCP frame is commanded to move forward 1 m (Fig. 6.6b) and then back to the initial frame (Fig. 6.6c). The controller leads to a completely different joint configuration when approaching the initial frame again. The second chart in Fig. 6.5 depicts a quadratic norm of selected null space subtask torques to allow direct comparison of the contributions. The return motion does not lead to the same subtask participation. For example, the upper body singularity avoidance is more crucial while moving forward to prevent the outstretched arm than it is while moving backward. In contrast, the avoidance of mechanical end stops only has a noteworthy effect during the backward motion (after 11 s). That complies well with the intuition of the observer when looking at the configurations of the robot in Fig. 6.6. Rollin' Justin is closer to its workspace boundaries in Fig. 6.6c than it is in Fig. 6.6a. The third chart in Fig. 6.5 depicts the Euclidean norms of the top priority tasks and the null space projection (projected subtasks from the second chart). It is noticeable that the collision avoidance only affects the behavior while moving backward. That is plausible: Since the arm is faster than the (inert) mobile base, a self-collision between the right

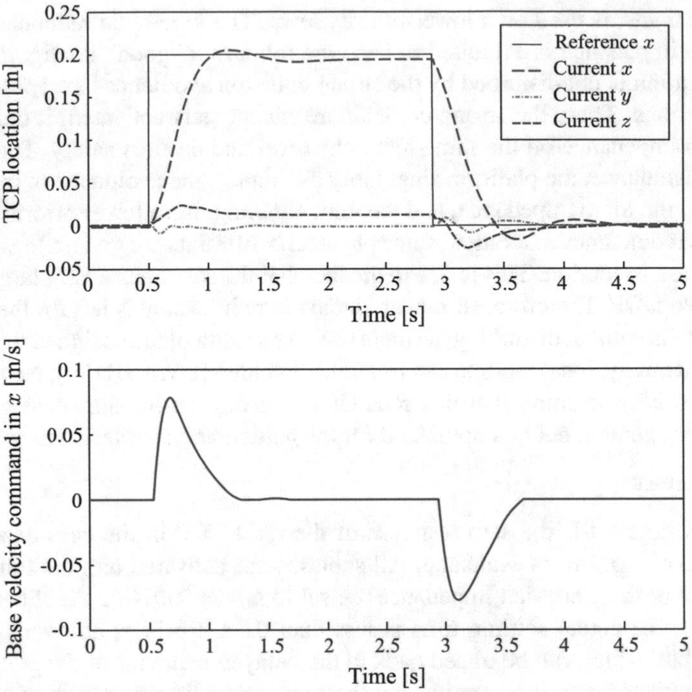

Fig. 6.4 *Experiment #1* The step response is evaluated for whole-body control with activated platform. The reference TCP location jumps $\Delta x = 0.2\,\mathrm{m}$ forward

hand and the torso has to be avoided while the platform is still accelerating. The last plot illustrates the base velocities which are the outputs of the platform admittance simulation. The significantly different configuration in Fig. 6.6c in comparison to the initial pose Fig. 6.6a is primarily caused by the asymmetrical commands for the mobile platform.

Physical Human–Robot Interaction

In experiment #3, the performance of physical human-robot interaction is analyzed. The user pushes the right TCP away from its desired position and orientation at about $t = 1\,\mathrm{s}$ and $t = 5\,\mathrm{s}$, see Fig. 6.7 (top chart). Thereupon, the mobile base tries to compensate for that error (bottom chart). This, in turn, leads to a null space motion w. r. t. the Cartesian impedance task. When releasing the TCP, the remaining platform velocity and the impedance induce a small overshoot before a steady state is reached again. That effect can be reduced by applying a higher stiffness to the TCP. An alternative would be to consider the platform velocity within the damping design of the Cartesian impedance. A deviation in the TCP orientation of almost 1 degree remains (middle chart). Two possible origins can be identified: On the one hand, the missing integrating component in the impedance controller (PD controller) prevents a zero steady-state error. On the other hand, the mobile base is designed to move only if a force threshold is exceeded. That avoids a permanent reorientation of the

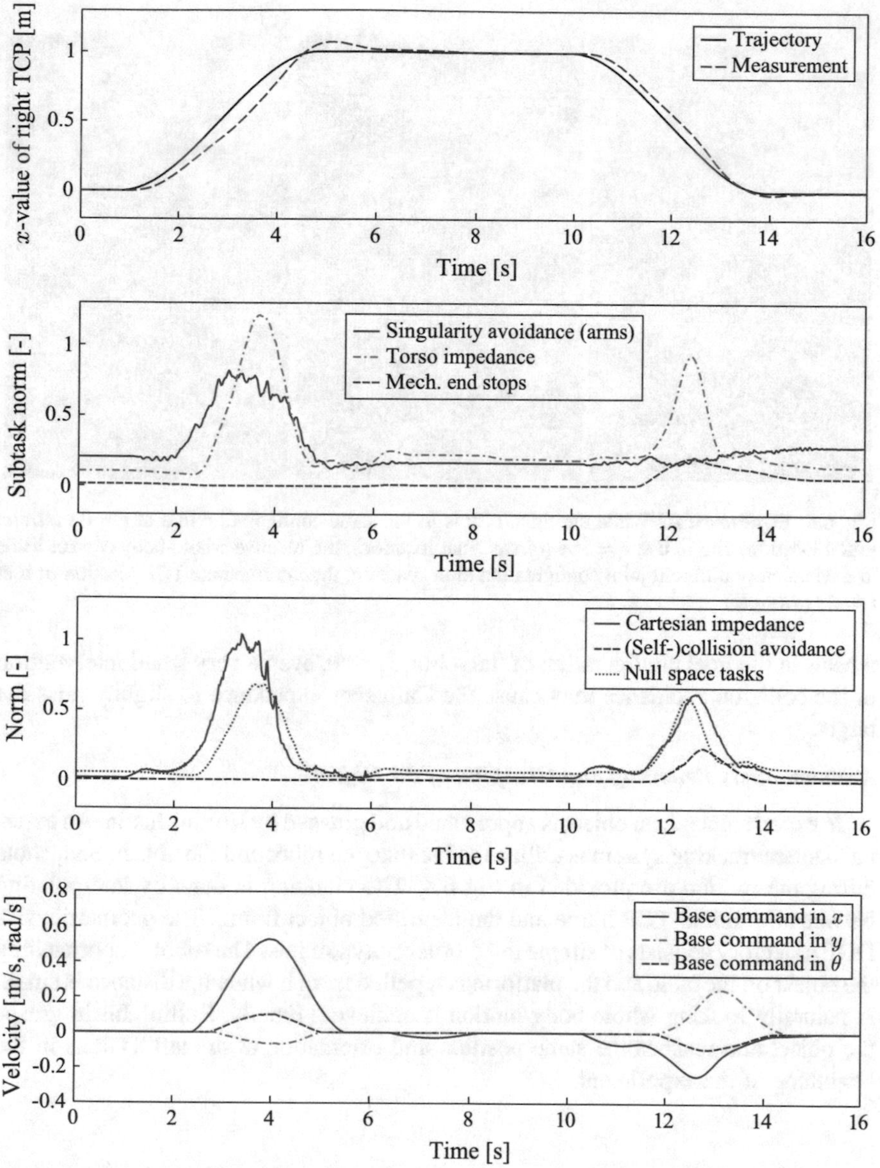

Fig. 6.5 *Experiment #2* A continuous trajectory for the right TCP is applied. The impedance stiffnesses are set to $k_{\text{tra}} = 500\,\text{N/m}$ and $k_{\text{rot}} = 100\,\text{Nm/rad}$. All damping ratios in the impedance law are set to $\xi = 0.7$

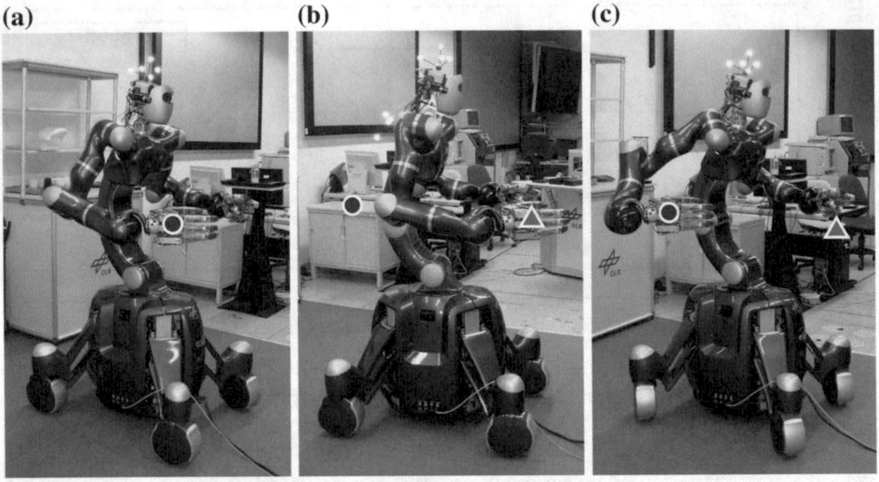

Fig. 6.6 *Experiment #2* While the right TCP is in the same configuration in **a** at $t = 0$ s (*circle*: initial location) and in **c** at $t = 16$ s (*circle*: final location), the reactive whole-body control leads to a completely different joint configuration after reaching the intermediate TCP location in **b** at $t = 8$ s (*triangle*)

wheels in the goal configuration of the robot. Hence, even a very small intervention of the collision avoidance may cause the Cartesian impedance to slightly miss the target.

Autonomously Reaching and Grasping of an Object

In experiment #4, an object is approached and grasped by Rollin' Justin. An external camera tracking system is utilized to localize the robot and the object. Snapshots during the motion are provided in Fig. 6.8. The planning is done by interpolating between the initial TCP frame and the identified object frame. The geometric, six-DOF trajectory consists of simple third-order polynomials. The robot is approaching the object on the table and the platform is repelled from it when the distance is small. A naturally looking whole-body motion is achieved. Finally, Rollin' Justin grasps the object and reaches the same position and orientation at the left TCP as in the beginning of the experiment.

6.3 Summary

In Sect. 6.1, the importance of the control tasks in a hierarchy was investigated to find a suitable order of priority. The whole-body impedance controller has been implemented in Sect. 6.2 on the basis of these considerations. The applied hierarchy utilized the results from the preceding chapters, e.g. the reactive control tasks from Chap. 3 as subgoals of the whole-body control, the null space projections for

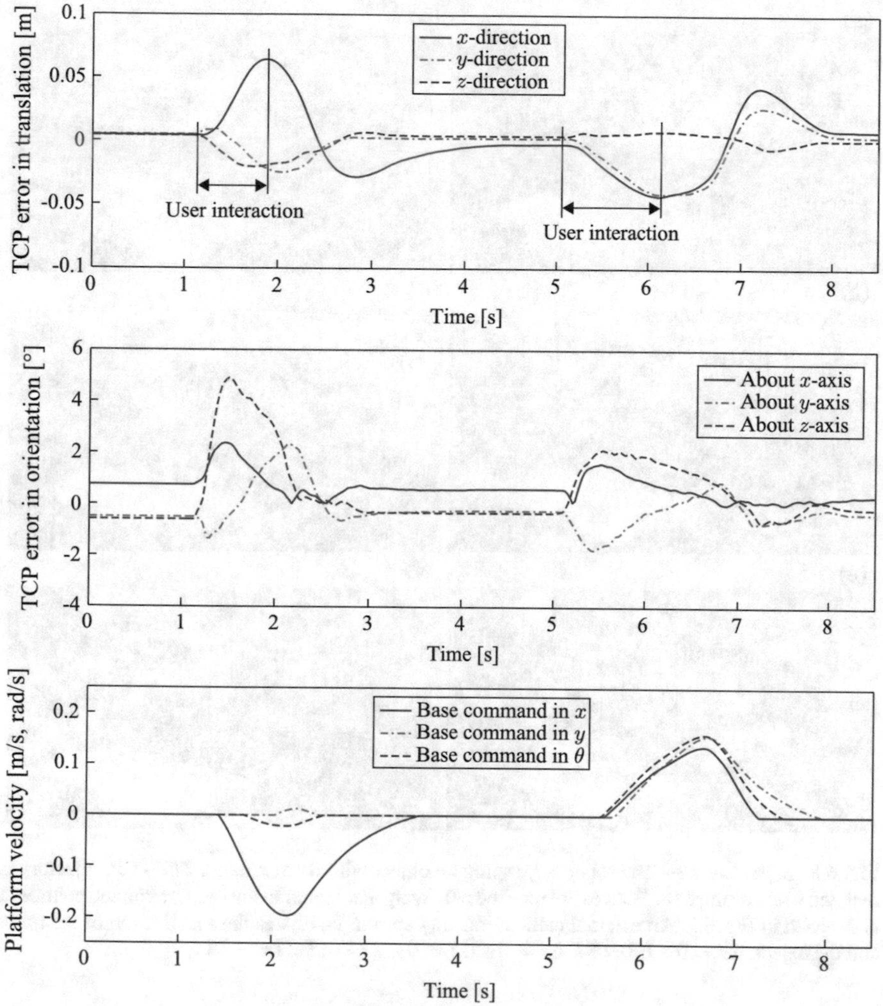

Fig. 6.7 *Experiment #3* Physical human–robot interaction is induced by the user. The translational and rotational deviations of the right TCP are shown as well as the platform commands of the whole-body controller

redundancy resolutions from Chap. 4, and the admittance coupling from Chap. 5 to provide the force-torque interface to the kinematically controlled mobile platform.

The experimental results have confirmed the theoretical findings and demonstrated the performance on a real system. The reactive nature of the approach could be observed in the experiments when the redundancy was resolved online according to the prescribed task hierarchy. Moreover, the evaluation of the physical human-robot interaction capabilities has clearly shown that the concept is particularly suitable for the operation in human environments due to the safe and soft contact behavior. Furthermore, the successful execution of an autonomous reaching-and-grasping-task

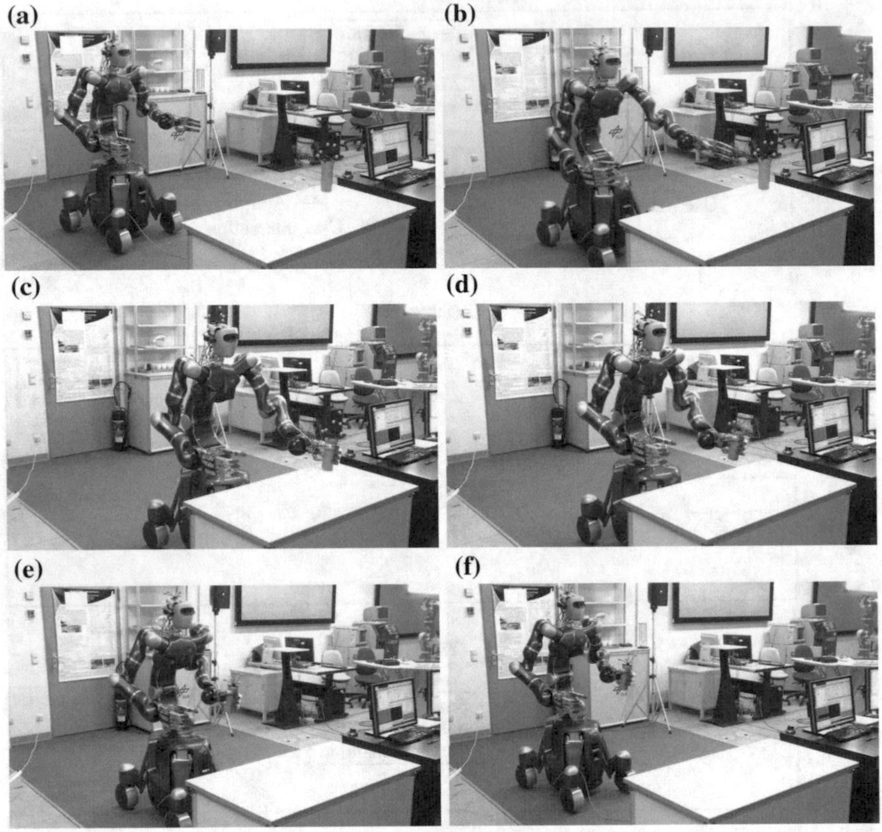

Fig. 6.8 *Experiment #4* The robot is grasping an object with the left hand. The TCP trajectory is realized while multiple objectives are reached reactively and simultaneously. The applied controller is depicted in Fig. 6.3. An external camera tracking system facilitates the localization of the robot and the object. **a** $t = 0$ s. **b** $t = 3$ s. **c** $t = 6$ s. **d** $t = 9$ s. **e** $t = 11$ s. **f** $t = 13$ s

has emphasized the advantage of the task definition in intuitive operational spaces: The complete robot with 51 actuated DOF was assigned to move to an object, grasp it, and move back by only defining a six-dimensional Cartesian trajectory for the end-effector.

Chapter 7
Integration of the Whole-Body Controller into a Higher-Level Framework

This chapter serves as an outlook to prospective challenges in robotics, where compliant whole-body control will act jointly with an *Artificial Intelligence* (AI). The field of service robotics, for example, puts high requirements on the systems due to the complexity of household chores: The environment is usually dynamic and unstructured, and a wide variety of tools with different contact properties exists. These aspects require an elaborate task planning, both from a logical and a geometric perspective. Moreover, the close cooperation of all modules in a robotic system is necessitated: A whole-body controller for soft physical contacts requires a proper parameterization, i.e. controller gains, a specified control task hierarchy, trajectories and goals, to perform the tasks. A non-deterministic, AI-based planner can provide these data while not necessarily being hard-real-time-capable itself. In case of local minima on the control level, the planner is able to reschedule to find feasible, global solutions. Other modules such as the vision system or the speech recognition may also be triggered. Figure 7.1 sketches how such a unified framework can combine and integrate different domains and how the subsystems are interconnected.

The successful execution of complex tasks requires the cooperation of two fundamentally different planning domains: *symbolic* planning and *geometric* planning. The symbolic planner can be described as the unit to generate logical schedules. Cleaning a window with a wiper could be symbolically described as: "locate the wiper", "grasp the wiper", "move to the window", "clean the window". The geometric planner is responsible for navigation, dynamics simulation, motion planning, and trajectory generation. The symbolic action "move to the window", for example, is geometrically interpreted to obtain a feasible platform trajectory to the window involving navigation within the world map of the robot. In the last years, there has been a steady progress in this new research field of *hybrid reasoning*, which combines symbolic planning and geometric planning.

In [WMR10], a hierarchical planning system was proposed which finds kinematic solutions for robotic manipulation. Results on discrete problems and pick-and-place tasks were implemented on PR2. In [KLP13], a combined approach for planning,

A. Dietrich, *Whole-Body Impedance Control of Wheeled Humanoid Robots*,
Springer Tracts in Advanced Robotics 116, DOI 10.1007/978-3-319-40557-5_7

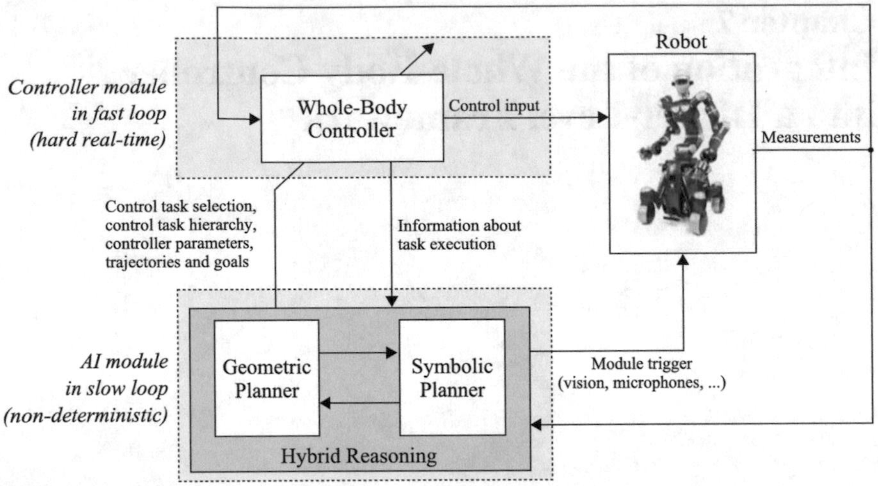

Fig. 7.1 Integrated framework for whole-body task execution. This sketch is an augmented version of Fig. 6.1 and based on the flow chart in [LDBAS16]

perception, state estimation, and action for mobile manipulation tasks is presented. Simulations and experiments on PR2 show that robustness and flexibility is given, even under modeling uncertainties. Reasoning methods were applied by Mösenlechner and Beetz [MB11] to enable the robot to find solutions such as where to place objects in a robust and flexible way. Dornhege et al. applied hybrid reasoning in [DGTN09]. While the symbolic planner was implemented via state-of-the-art methods, the geometric part was realized by means of a probabilistic roadmap algorithm. The focus of this work was to generate collision-free trajectories.

A unified framework utilizing hybrid reasoning on Rollin' Justin has been presented in [LDS+14, LDBAS16], which implements the structure in Fig. 7.1, yet missing the communication channel from controller to AI-based planner. The central point is the parameterization of the controller by the planner, i.e. the communication channel from the planner to the whole-body controller. Section 7.1 addresses this part from a control point of view. The other interconnection, i.e. the feedback from the controller to the AI-based planner, is treated in Sect. 7.2 and serves as an outlook to ongoing research activities on integrated frameworks. Three realistic service robot tasks performed by Rollin' Justin are shown in Sect. 7.3. The summary in Sect. 7.4 concludes this chapter and highlights the potential of an integrated framework as the one depicted in Fig. 7.1.

7.1 Intelligent Parameterization of the Whole-Body Controller

The AI module selects the required control tasks and establishes a hierarchy among them, depending on the requirements of the considered application. Moreover, it parameterizes the control tasks and commands trajectories and goals. A concept for the automated parameterization will be described in the following.

Relying on an object database makes it possible for the symbolic and the geometric planner to parameterize all control tasks w. r. t. the requirements of the environment and the involved objects. An example is given in Table 7.1, where three different tools are specified in terms of their handling. At the example of a window wiper (center column), one can see that specific Cartesian force limits along the x-axis of the tool are defined, and the stiffness for the rotation about the x-axis is set very high. These exemplary values represent the typical use of such tools, i.e. how a human would actually perform the respective force-sensitive household chore [UASvdS04]. Note that Table 7.1 only shows a small subset of all possible parameters. In many cases, default values are used, e.g. for the parameterization of the self-collision avoidance or the thresholds for the arm singularity avoidance. The geometric planner requires information from the object database. The window, which is also contained in the database, has specific dimensions and properties and has therefore a direct influence on the trajectories to be computed and the proper use of the corresponding tool.

The parameterizations of the three tools in Table 7.1 are experimentally validated in Sect. 7.3.

7.2 Communication Channel Between Controller and Planner

The communication channel from the controller to the AI module in Fig. 7.1 has been introduced to account for a high-level supervision and reaction, which is required when the controller faces problems that it cannot solve itself (e.g. undesired local minima or inconsistent trajectories due to altered environmental conditions). This aspect is currently topic of intensive research.

The basic idea is to feedback preprocessed data from the control level to the AI module, which contains valuable information about the task execution. Besides obvious choices such as the Cartesian error at the TCP, this might also include physical values such as the momentum (to detect contacts), frequency information (to assess the performance in cyclic cleaning motions), or the energy consumption (to evaluate the efficiency of the applied method). Moreover, it is reasonable to define storage functions of the subtask controllers at different hierarchy levels to identify local minima and conflicting tasks.

Table 7.1 Controller parameterization for the service tasks performed by Rollin' Justin

	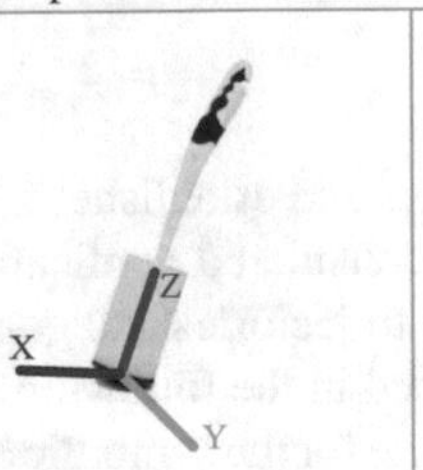	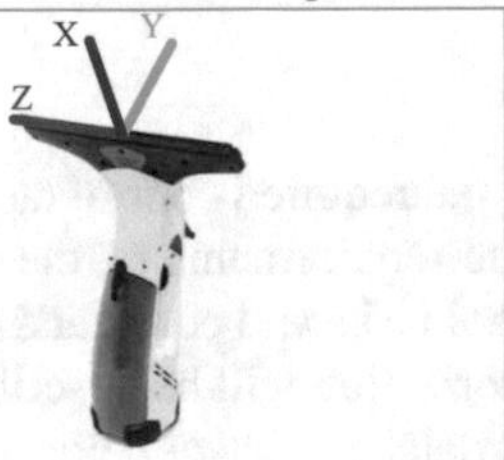	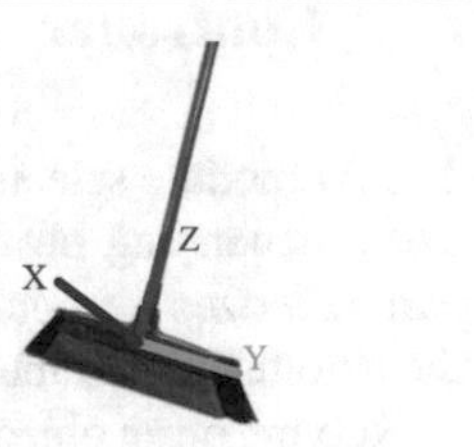
Robot task	"*Clean the mug!*"	"*Wipe the window!*"	"*Sweep the floor!*"
Actuators	Both arms, torso	One arm, torso, base	Both arms, torso, base
Task hierarchy	Cartesian impedance, joint impedance, end stop avoidance, singularity avoidance	Cartesian impedance, self-collision avoidance, singularity avoidance, joint impedance, end stop avoidance	Cartesian impedance, self-collision avoidance, joint impedance, end stop avoidance, singularity avoidance
Transl. stiffness	(400, 400, 800) N/m	(100, 500, 1000) N/m	(1000, 500, 300) N/m
Rot. stiffness	(30, 30, 60) Nm/rad	(500, 10, 10) Nm/rad	(200, 10, 500) Nm/rad
Force limits	$(\pm 20, \pm 20, \pm 20)$ N	$(-\infty/+10, \pm\infty, \pm\infty)$ N	$(\pm\infty, \pm\infty, -10/+\infty)$ N

The stiffnesses and force limits refer to the Cartesian coordinates of the tools (in the order: x, y, z) as depicted in the respective photographs

By means of various techniques such as learning algorithms, the collected information can then be interpreted (e.g. stop task, adapt symbolic plan, replan trajectory), and an appropriate response is generated and given back to the whole-body controller.

7.3 Real-World Applications for a Service Robot

Three different household chores have been experimentally validated based on the objects in Table 7.1. The Cartesian trajectories of the left and the right TCP are provided by the higher-level instance and are depicted in the left diagrams in Figs. 7.2, 7.3, and 7.4. The reference values of the tools are deliberately placed *behind* the surface to be cleaned (shaded areas in the snapshots) to account for modeling and vision errors in the estimation of the surface location. Accordingly, there results a large but deliberate difference between the commanded TCP positions and the measured ones as can be seen in the plots on the left.

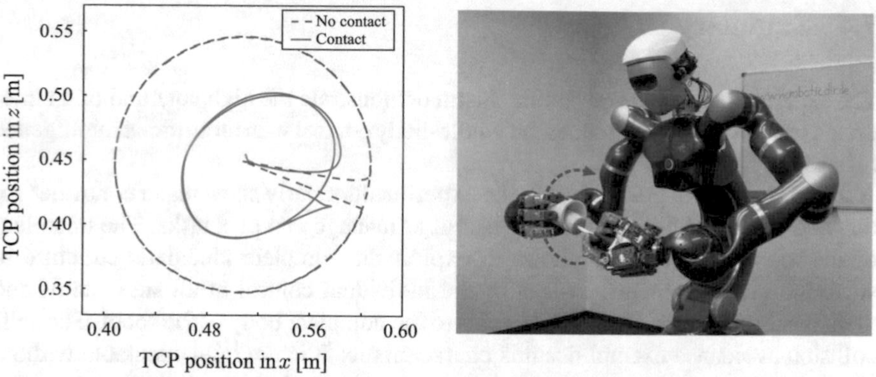

Fig. 7.2 Application: "cleaning a mug"

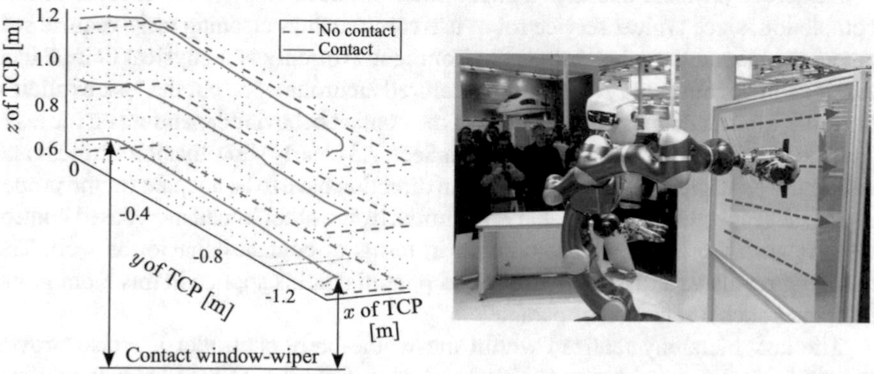

Fig. 7.3 Application: "wiping a window"

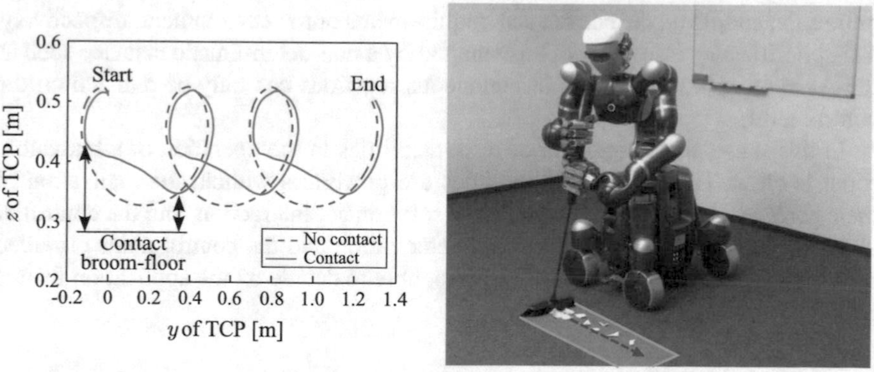

Fig. 7.4 Application: "sweeping the floor"

7.4 Summary

The experimental results on Rollin' Justin demonstrate the high potential of an integrated framework that combines the whole-body control with an artificial intelligence instance.

From the control point of view, the experiments clearly show that a controller for the whole body of the robot is very useful to manage complex tasks. The treatment of the robot in its entirety allows to exploit the complete kinematic structure to solve the given problems. Instead of the individual control of all subsystems and their synchronization, the subtasks refer to the complete body of the robot. The self-collision avoidance exemplifies this characteristic: It is not implementable without instantaneous access to all actuated joints in the whole body. This property also applies to the other tasks in the control hierarchy.

Dexterous physical interaction capabilities are necessary for the successful task completion, since typical service robot use cases such as cleaning tasks require soft contacts between the robot and its environment. Not only is a physical compliance essential in dynamic and partially unstructured environments, but the task explicitly demands a particular parameterization of the contact behavior depending on the task-specific requirements. The applications in Sect. 7.3 clearly show that the stiffness and maximum contact forces in each Cartesian direction have to be adapted for the proper use of the respective tool, cf. Table 7.1. Similarly, the other impedance-based control tasks require a proper parameterization in terms of contact behavior as well. The theory of repulsive and attractive artificial potential fields applied in this monograph facilitates such a selection of parameters.

The task hierarchy realized within the whole-body controller has also proven successful in the experiments. Higher-level goals from the AI-based planner claim a control task prioritization, since the robot usually has "more important" and "less important" objectives during the task execution. The order of priority may even change online, depending on the current task requirements or the environment, respectively. This prioritization issue cannot be managed by a non-deterministic instance such as the AI in an appropriate way. Instantaneous reactions can only be realized on the control level.

In this sense, the assignment of responsibilities in the operation of a humanoid robot is clear: The higher-level modules are powerless without the realization of their goals by a controller with skills for soft contact interaction, and the controller alone is blind without the proper parameterization and the control goals provided by a higher-level planner which is, in turn, able to decide on the application from a global perspective.

Chapter 8
Summary

The steady progress in humanoid robotics has recently generated impressive systems with a large number of actuated degrees of freedom. Consequently, that gave a strong impetus to research activities related to the whole-body control of this newly available class of mobile robots. The demand arises from various fields of application as sketched in Fig. 8.1. Industrial use cases are very relevant since robotic systems can be employed to perform dangerous and harmful tasks, or they could assist factory workers and cooperate with them. Another field concerns hazardous environments such as in space missions, deep see exploration, or disasters in nuclear reactors, e.g. Chernobyl or Fukushima Daiichi. Further applications are encountered in the emerging field of service robotics. Even simple household chores such as washing the dishes require a lot from the robotic systems because the considered environment is often dynamic, unstructured, and difficult to predict due to the presence of human beings.

In order to efficiently and robustly execute tasks without endangering the humans in the workspace, sophisticated whole-body controllers must be employed. Compliant physical interaction with the environment of the robot is one of the key aspects here. Not only is this feature important for safety reasons, but also for the effective task execution. Examples are cleaning tasks, where the contact behavior is predefined, or physical human–robot interaction, where the robotic system is required to be soft in contact with the human. The goal of this monograph was to present solutions for compliant whole-body control. The extensions of the well-established concept of impedance control particularly addressed the aspect of whole-body control for a hierarchically arranged set of tasks and objectives. While most of the presented techniques were implementable on generic robots, several methods have been particularly developed for the use on wheeled systems. In the following paragraphs, an overview of the content and the contributions in the chapters is given.

Multi-objective control only makes sense if a wide range of useful objectives is available for the parameterization of the task hierarchy. Depending on the requirements of the envisioned application, a suitable set of tasks must be chosen to satisfy them. Chapter 3 addressed this topic and extended the standard repertoire of commonly available control objectives by three valuable tasks. That included a reactive

A. Dietrich, *Whole-Body Impedance Control of Wheeled Humanoid Robots*,
Springer Tracts in Advanced Robotics 116, DOI 10.1007/978-3-319-40557-5_8

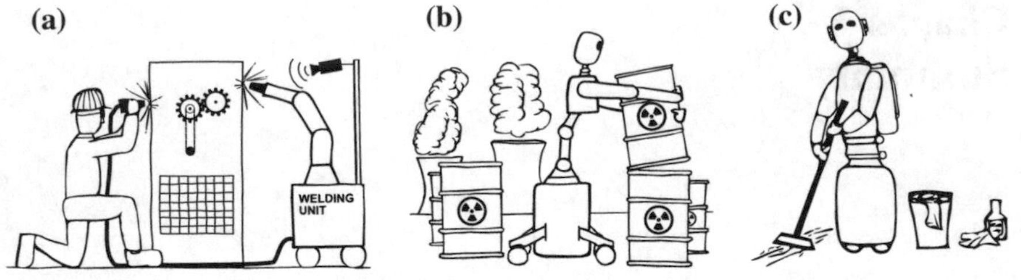

Fig. 8.1 Real-world applications for mobile whole-body control. **a** Industrial applications. **b** Operation in dangerous environments. **c** Household robots

self-collision avoidance algorithm, which uses artificial potential fields to repel body parts from each other. Multi-DOF structures of modern humanoid robots necessitate the integration of such a safety feature because of the large number of potentially colliding links. Apart from the self-protection of the robot, also humans in its workspace are safer due to the avoidance of harmful clamping. Furthermore, the algorithm comprises a configuration-dependent damping design, which realizes a desired mass-spring-damper behavior in the collision directions. The second reactive task introduced in Chap. 3 was a singularity avoidance technique for nonholonomic, wheeled platforms. During motions, the instantaneous center of rotation may come close to the steering axis of a wheel, requiring an infinite steering rate to comply with the nonholonomic rolling constraints. The proposed algorithm avoids these critical situations by repelling the instantaneous center of rotation from the steering axis. The purpose of the third reactive task in Chap. 3 was to exploit the full torso workspace of a robot with kinematically coupled torso joints. Both kinematic constraints (joint limits) and dynamic constraints (torque limits) restrict the reachable workspace. The method allows to use the maximum feasible range of the torso by online adaption of repulsive forces to avoid prohibited configurations and motions.

Chapter 4 addressed the concept of null space projections to realize a task hierarchy among the chosen objectives for the current application. The first contribution of this chapter was the review of all relevant null space projectors for the redundancy resolution in torque control. That included the analysis of different types of hierarchy structures, namely the successive and the augmented approaches. Moreover, different hierarchy consistencies were investigated, e.g., the popular dynamically consistent solution and the statically consistent one. The theory of dynamic consistency was extended and yielded a deeper understanding of the dynamic decoupling of priority levels in general. Additionally, a new kind of null space projection was introduced, which was named the stiffness-consistent method. If parallel elastic actuators are used in a robot, e.g., when mechanical springs are mounted in parallel to the motors to counterbalance the gravitational loads, this new null space projector makes it possible to still realize the desired task hierarchy by taking these additional passive elements into account. Without knowledge of the stiffness of the mechanical springs in the projector calculation, the springs and the controller would inevitably compete with each other. All projectors have been compared in simulations and experiments.

Among others, it was shown that the theoretically superior, classical approaches lose their dominating position on real hardware. The second contribution of Chap. 4 was a solution to the problems of unilateral constraints in the task hierarchy, task singularities, and dynamic modifications in the order of priority. The approach has been theoretically and experimentally validated and has proven a powerful tool in whole-body control to deal with these major issues.

A thorough study of the literature on whole-body control quickly reveals a key problem of the state of the art: Many approaches are proposed and partially implemented on real hardware, but stability analyses are extremely rare. Nevertheless, the operation of robots in human environments requires a maximum level of safety. Therefore, it is essential to understand the stability properties and verify that safety is actually given. Chapter 5 addressed that issue and provided the stability analyses. The first part contributed to the field of mobility. As the concept of impedance cannot be straightforwardly applied to a kinematically controlled mobile base with torque-controlled upper body, an admittance interface to the platform was proposed to integrate it into the overall whole-body impedance framework. But this interconnection also resulted in modified dynamic equations and closed-loop instability. For that reason, a model-based compensation term was designed to stabilize the system. Asymptotic stability of the equilibrium and passivity features were demonstrated. The obtained stabilizing controller finally yielded a whole-body impedance behavior for wheeled robots with position- or velocity-controlled mobile platform. The second part in Chap. 5 contributed to multi-objective control. The majority of the state-of-the-art methods uses the so-called *Operational Space Formulation*. However, it is well known that this classical controller only gives a proof of stability for the main task, while the complete null space stability is unclear. Here, a stability analysis for the complete robot with an arbitrary number of priority levels has been performed that finally led to the conclusion of asymptotic stability of the desired, hierarchy-consistent equilibrium. A new formulation of the dynamic equations was proposed that takes the strict task hierarchy into account. This new representation and the theory of conditional stability were the key aspects for the stability analysis. Furthermore, compared to the classical approaches, the control law does not require the problematic feedback of external forces and torques because it preserves the natural inertia of the robot. The considered system is generic so that the results can be applied to any robotic system with force or torque input such as a legged humanoid or an aerial vehicle with a manipulator attached to it. Therefore, it is also possible to combine the two proofs of stability of Chap. 5 in one system: a wheeled mobile manipulator with impedance-based task hierarchy for any given number of priority levels.

The experiments conducted in Chap. 5 emphasized the practical importance of the theoretical proofs of stability. A demonstrative example was given in Sect. 5.1: Without the compensation term, which resulted from the formal stability analysis, the robot operated properly in most cases. However, slightly modifying the controller gains suddenly led to instability. When using the compensation term in the controller, stability could be shown for *all* parameterizations, both theoretically and experimentally. In a nutshell, Chap. 5 confirmed Kurt Lewin's (1890–1947) famous quote "*There is nothing so practical as a good theory*" and highlighted its relevance for robotics.

Besides the theoretical contributions in this book, the experimental evaluations occupied an important role. Chapter 6 was put in place to investigate the reactive whole-body coordination using the results from the preceding chapters. Here, the focus was to evaluate how a reactive whole-body controller with multiple objectives resolves the redundancy from a priority-based perspective.

The proposed whole-body impedance control framework is the main result of this monograph, and it contributes to the state of the art in various fields such as control design, stability theory, and the experimental validation of concepts. In order to make the approach widely usable in the future, it is necessary to consider it also as an integrated component in a framework of larger scale. The operation of a mobile humanoid robot in a domestic environment, for example, requires the close cooperation of different domains. Among others, that includes the interpretation of an abstract, user-given task from a logical point of view, image processing based on camera data, geometric planning of reasonable trajectories, task-specific parameterization of the whole-body controller, and the control itself. In this respect, the final Chap. 7 served two purposes. First, it sketched the integration into such an overall framework. Three complex household chores have been selected for the evaluation on the humanoid robot Rollin' Justin which required both an artificial intelligence module and a whole-body controller for compliant physical interaction and task execution: cleaning a mug, wiping a window, and sweeping the floor with a broom. The experimental results have confirmed the great potential of the approach in terms of real-world applications. The integration of all necessary components including a whole-body controller to solve such complex household chores clearly distinguishes the proposed approach from state-of-the-art solutions.

The second purpose of Chap. 7 was to serve as an outlook for future research activities in this field, because the investigation of several aspects of integrated frameworks is still in its early stages in the robotics community. It is of high relevance to stay tuned to the *fusion* of all involved research areas in robotics, so that the goal of an autonomous humanoid robot solving relevant everyday tasks can be reached one day.

Appendix A
Workspace of the Torso of Rollin' Justin

A.1 Kinematic Constraints

The kinematic workspace boundaries in Fig. 3.15b can be determined by substituting the joint values in (3.42) by the mechanical limits in Table 3.2. The obtained analytical expressions represent circles in the X-Z-plane. Their centers and radii are listed in Table A.1. The arc numbers correspond to the ones in Fig. 3.15b.

A.2 Dynamic Constraints

The dynamic constraints due to limitations of the third torso joint torque can be described by circles in the X-Z-plane. These boundaries can be derived by combining (3.47), (3.49), (3.51), and (3.52). They are illustrated in Fig. (3.16) and mathematically described in Table A.2.

Table A.1 Centers and radii of the arcs defining the kinematic workspace boundaries of the torso of Rollin' Justin

Arc	Constraint	Center	Radius
1	$q_{t,2} = -90°$	$(-l_{t,2}, 0)$	$l_{t,3}$
2	$q_{t,3} = 135°$	$(0, 0)$	$\sqrt{l_{t,2}^2 + l_{t,3}^2 - \sqrt{2} l_{t,2} l_{t,3}}$
3	$q_{t,3} = 135° - q_{t,2}$	$(l_{t,3}/\sqrt{2}, -l_{t,3}/\sqrt{2})$	$l_{t,2}$
4	$q_{t,2} = 90°$	$(l_{t,2}, 0)$	$l_{t,3}$
5	$q_{t,3} = 0°$	$(0, 0)$	$l_{t,2} + l_{t,3}$
6	$q_{t,3} = -q_{t,2}$	$(0, l_{t,3})$	$l_{t,2}$

A. Dietrich, *Whole-Body Impedance Control of Wheeled Humanoid Robots*, Springer Tracts in Advanced Robotics 116, DOI 10.1007/978-3-319-40557-5

Table A.2 Centers and types of the circular, dynamic workspace constraints of the torso of Rollin' Justin due to torque limitations of the third torso joint

$\tau_{t,3}$	x-**value of center**	z-**value of center**	**Workspace**
$\tau_{t,3,\min}$	$l_{t,3}\sin\left(\alpha+\arccos\left(\frac{-\tau_{t,3,\min}}{F_{t,\text{load}}l_{t,3}}\right)\right)$	$l_{t,3}\cos\left(\alpha+\arccos\left(\frac{-\tau_{t,3,\min}}{F_{t,\text{load}}l_{t,3}}\right)\right)$	Inside
$\tau_{t,3,\min}$	$l_{t,3}\sin\left(\alpha-\arccos\left(\frac{-\tau_{t,3,\min}}{F_{t,\text{load}}l_{t,3}}\right)\right)$	$l_{t,3}\cos\left(\alpha-\arccos\left(\frac{-\tau_{t,3,\min}}{F_{t,\text{load}}l_{t,3}}\right)\right)$	Outside
$\tau_{t,3,\max}$	$l_{t,3}\sin\left(\alpha+\arccos\left(\frac{-\tau_{t,3,\max}}{F_{t,\text{load}}l_{t,3}}\right)\right)$	$l_{t,3}\cos\left(\alpha+\arccos\left(\frac{-\tau_{t,3,\max}}{F_{t,\text{load}}l_{t,3}}\right)\right)$	Outside
$\tau_{t,3,\max}$	$l_{t,3}\sin\left(\alpha-\arccos\left(\frac{-\tau_{t,3,\max}}{F_{t,\text{load}}l_{t,3}}\right)\right)$	$l_{t,3}\cos\left(\alpha-\arccos\left(\frac{-\tau_{t,3,\max}}{F_{t,\text{load}}l_{t,3}}\right)\right)$	Inside

Appendix B
Null Space Definitions and Proofs

For the sake of simplicity, the dependencies on $\boldsymbol{q}$ are omitted in the following notations.

B.1 Representations of Null Space Projectors

The generic, non-singular Jacobian matrix $\boldsymbol{J} \in \mathbb{R}^{m \times n}$ for $m < n$ with its singular value decomposition

$$\boldsymbol{J} = \boldsymbol{U}\boldsymbol{S}\boldsymbol{V}^T , \tag{B.1}$$

$$\boldsymbol{V} = \left(\boldsymbol{X}^T , \boldsymbol{Y}^T\right) , \tag{B.2}$$

$$\boldsymbol{S} = (\boldsymbol{\Sigma} , \boldsymbol{0}) , \tag{B.3}$$

is used. The matrices $\boldsymbol{U} \in \mathbb{R}^{m \times m}$ and $\boldsymbol{V} \in \mathbb{R}^{n \times n}$ are orthonormal and the rectangular diagonal matrix $\boldsymbol{S} \in \mathbb{R}^{m \times n}$ contains the singular values σ_1 to σ_m in its submatrix $\boldsymbol{\Sigma} \in \mathbb{R}^{m \times m}$. The range space of $\boldsymbol{J}$ is defined by $\boldsymbol{X} \in \mathbb{R}^{m \times n}$ and its null space is determined by $\boldsymbol{Y} \in \mathbb{R}^{(n-m) \times n}$. The projector onto the null space of $\boldsymbol{J}$ is denoted $\boldsymbol{N} \in \mathbb{R}^{n \times n}$.

Theorem B.1 *A null space projection onto the null space of a full-row-rank Jacobian matrix* $\boldsymbol{J} \in \mathbb{R}^{m \times n}$ *for* $m < n$ *is invariant to the singular values of* $\boldsymbol{J}$*. The range space* $\boldsymbol{X} \in \mathbb{R}^{m \times n}$ *of* $\boldsymbol{J}$ *is sufficient to compute the projector. The null space projector definition*

$$\boldsymbol{N} = \boldsymbol{I} - \boldsymbol{J}^T (\boldsymbol{J}^{W+})^T \tag{B.4}$$

$$= \boldsymbol{I} - \boldsymbol{X}^T (\boldsymbol{X}^{W+})^T \tag{B.5}$$

holds for any non-singular weighting matrix $\boldsymbol{W} \in \mathbb{R}^{n \times n}$.

A. Dietrich, *Whole-Body Impedance Control of Wheeled Humanoid Robots*,
Springer Tracts in Advanced Robotics 116, DOI 10.1007/978-3-319-40557-5

Proof Equation (B.4) is the standard definition of the null space projector [Kha87, SS91] and (B.5) refers to a representation which is independent of the singular values of $\boldsymbol{J}$. Expanding both formulas according to the weighted pseudoinversion (4.15) and cancelling the identity matrix yields

$$\boldsymbol{J}^T \left(\boldsymbol{J}\boldsymbol{W}^{-1}\boldsymbol{J}^T\right)^{-T} \boldsymbol{J}\boldsymbol{W}^{-T} = \boldsymbol{X}^T \left(\boldsymbol{X}\boldsymbol{W}^{-1}\boldsymbol{X}^T\right)^{-T} \boldsymbol{X}\boldsymbol{W}^{-T} \tag{B.6}$$

Now $\boldsymbol{W}^{-T}$ can be eliminated, and the multiplication by $\boldsymbol{V}^T$ (from the left) and $\boldsymbol{V}$ (from the right) delivers

$$\begin{aligned}
\boldsymbol{V}^T\boldsymbol{V}\boldsymbol{S}^T\boldsymbol{U}^T\left[\boldsymbol{U}\boldsymbol{S}\boldsymbol{V}^T\boldsymbol{W}^{-1}\boldsymbol{V}\boldsymbol{S}^T\boldsymbol{U}^T\right]^{-T}\boldsymbol{U}\boldsymbol{S}\boldsymbol{V}^T\boldsymbol{V} &= \boldsymbol{V}^T\boldsymbol{X}^T\left[\boldsymbol{X}\boldsymbol{W}^{-1}\boldsymbol{X}^T\right]^{-T}\boldsymbol{X}\boldsymbol{V}\\
\begin{bmatrix}\boldsymbol{\Sigma}^T\boldsymbol{U}^T\\ \boldsymbol{0}\end{bmatrix}\left[[\boldsymbol{U}\boldsymbol{\Sigma},\boldsymbol{0}]\,\boldsymbol{V}^T\boldsymbol{W}^{-1}\boldsymbol{V}\begin{bmatrix}\boldsymbol{\Sigma}^T\boldsymbol{U}^T\\ \boldsymbol{0}\end{bmatrix}\right]^{-T}[\boldsymbol{U}\boldsymbol{\Sigma},\boldsymbol{0}] &= \begin{bmatrix}\boldsymbol{I}\\ \boldsymbol{0}\end{bmatrix}\left(\boldsymbol{X}\boldsymbol{W}^{-1}\boldsymbol{X}^T\right)^{-T}[\boldsymbol{I},\boldsymbol{0}]\\
\begin{bmatrix}\boldsymbol{\Sigma}^T\boldsymbol{U}^T\\ \boldsymbol{0}\end{bmatrix}\left[(\boldsymbol{U}\boldsymbol{\Sigma})(\boldsymbol{X}\boldsymbol{W}^{-1}\boldsymbol{X}^T)(\boldsymbol{\Sigma}^T\boldsymbol{U}^T)\right]^{-T}[\boldsymbol{U}\boldsymbol{\Sigma},\boldsymbol{0}] &= \begin{bmatrix}\boldsymbol{I}\\ \boldsymbol{0}\end{bmatrix}\left(\boldsymbol{X}\boldsymbol{W}^{-1}\boldsymbol{X}^T\right)^{-T}[\boldsymbol{I},\boldsymbol{0}]\\
\begin{bmatrix}\boldsymbol{\Sigma}^T\boldsymbol{U}^T\\ \boldsymbol{0}\end{bmatrix}(\boldsymbol{\Sigma}^T\boldsymbol{U}^T)^{-1}(\boldsymbol{X}\boldsymbol{W}^{-T}\boldsymbol{X}^T)^{-1}(\boldsymbol{U}\boldsymbol{\Sigma})^{-1}[\boldsymbol{U}\boldsymbol{\Sigma},\boldsymbol{0}] &= \begin{bmatrix}\boldsymbol{I}\\ \boldsymbol{0}\end{bmatrix}\left(\boldsymbol{X}\boldsymbol{W}^{-T}\boldsymbol{X}^T\right)^{-1}[\boldsymbol{I},\boldsymbol{0}]\\
\begin{bmatrix}\left(\boldsymbol{X}\boldsymbol{W}^{-T}\boldsymbol{X}^T\right)^{-1} & \boldsymbol{0}\\ \boldsymbol{0} & \boldsymbol{0}\end{bmatrix} &= \begin{bmatrix}\left(\boldsymbol{X}\boldsymbol{W}^{-T}\boldsymbol{X}^T\right)^{-1} & \boldsymbol{0}\\ \boldsymbol{0} & \boldsymbol{0}\end{bmatrix},
\end{aligned}$$

which finally proves Theorem B.1. □

Theorem B.2 *A projector onto the null space of a full-row-rank Jacobian matrix* $\boldsymbol{J} \in \mathbb{R}^{m\times n}$ *for* $m < n$ *can be computed by sparing the range space* $\boldsymbol{X} \in \mathbb{R}^{m\times n}$ *of* $\boldsymbol{J}$ *and only utilizing its null space* $\boldsymbol{Y} \in \mathbb{R}^{(n-m)\times n}$. *The identity*

$$\boldsymbol{N} = \boldsymbol{I} - \boldsymbol{X}^T(\boldsymbol{X}^{\boldsymbol{W}+})^T \tag{B.7}$$

$$= \boldsymbol{W}^T\boldsymbol{Y}^T(\boldsymbol{Y}\boldsymbol{W}^T\boldsymbol{Y}^T)^{-1}\boldsymbol{Y} \tag{B.8}$$

holds for any non-singular weighting matrix $\boldsymbol{W} \in \mathbb{R}^{n\times n}$.

Proof Equation (B.7) is the (already proven) null space projector according to Theorem B.2, (B.8) is a projector representation without utilization of the range space of $\boldsymbol{J}$. Equating and reorganizing both formulas delivers

$$\boldsymbol{I} = \boldsymbol{X}^T\left(\boldsymbol{X}\boldsymbol{W}^{-1}\boldsymbol{X}^T\right)^{-T}\boldsymbol{X}\boldsymbol{W}^{-T} + \boldsymbol{W}^T\boldsymbol{Y}^T\left(\boldsymbol{Y}\boldsymbol{W}^T\boldsymbol{Y}^T\right)^{-1}\boldsymbol{Y}\,. \tag{B.9}$$

The multiplication by $\boldsymbol{V}^T$ (from the left) and $\boldsymbol{V}$ (from the right) leads to

$$\begin{aligned}
\boldsymbol{I} &= \begin{bmatrix}\boldsymbol{I}\\ \boldsymbol{0}\end{bmatrix}\left(\boldsymbol{X}\boldsymbol{W}^{-T}\boldsymbol{X}^T\right)^{-1}[\boldsymbol{X}\boldsymbol{W}^{-T}\boldsymbol{X}^T,\boldsymbol{X}\boldsymbol{W}^{-T}\boldsymbol{Y}^T] + \begin{bmatrix}\boldsymbol{X}\boldsymbol{W}^T\boldsymbol{Y}^T\\ \boldsymbol{Y}\boldsymbol{W}^T\boldsymbol{Y}^T\end{bmatrix}\left(\boldsymbol{Y}\boldsymbol{W}^T\boldsymbol{Y}^T\right)^{-1}[\boldsymbol{0},\boldsymbol{I}]\\
&= \begin{bmatrix}\boldsymbol{I} & \left(\boldsymbol{X}\boldsymbol{W}^{-T}\boldsymbol{X}^T\right)^{-1}\left(\boldsymbol{X}\boldsymbol{W}^{-T}\boldsymbol{Y}^T\right)\\ \boldsymbol{0} & \boldsymbol{0}\end{bmatrix} + \begin{bmatrix}\boldsymbol{0} & \left(\boldsymbol{X}\boldsymbol{W}^T\boldsymbol{Y}^T\right)\left(\boldsymbol{Y}\boldsymbol{W}^T\boldsymbol{Y}^T\right)^{-1}\\ \boldsymbol{0} & \boldsymbol{I}\end{bmatrix}.
\end{aligned} \tag{B.10}$$

Equation (B.10) is true if the upper right element is zero. Therefore, one has to verify

$$\left(\boldsymbol{X}\boldsymbol{W}^{-T}\boldsymbol{X}^T\right)^{-1}\left(\boldsymbol{X}\boldsymbol{W}^{-T}\boldsymbol{Y}^T\right)+\left(\boldsymbol{X}\boldsymbol{W}^T\boldsymbol{Y}^T\right)\left(\boldsymbol{Y}\boldsymbol{W}^T\boldsymbol{Y}^T\right)^{-1}=\boldsymbol{0}\,. \tag{B.11}$$

That can be achieved by multiplying $\boldsymbol{X}\boldsymbol{W}^{-T}\boldsymbol{X}^T$ (from the left) and $\boldsymbol{Y}\boldsymbol{W}^T\boldsymbol{Y}^T$ (from the right).

$$\begin{aligned}\boldsymbol{X}\boldsymbol{W}^{-T}\boldsymbol{Y}^T\boldsymbol{Y}\boldsymbol{W}^T\boldsymbol{Y}^T+\boldsymbol{X}\boldsymbol{W}^{-T}\boldsymbol{X}^T\boldsymbol{X}\boldsymbol{W}^T\boldsymbol{Y}^T&=\boldsymbol{0}\\ \boldsymbol{X}\boldsymbol{W}^{-T}\left(\boldsymbol{Y}^T\boldsymbol{Y}+\boldsymbol{X}^T\boldsymbol{X}\right)\boldsymbol{W}^T\boldsymbol{Y}^T&=\boldsymbol{0}\\ \boldsymbol{X}\boldsymbol{Y}^T&=\boldsymbol{0}\\ \boldsymbol{0}&=\boldsymbol{0}\end{aligned}$$

Note that $\boldsymbol{Y}^T\boldsymbol{Y}+\boldsymbol{X}^T\boldsymbol{X}=\boldsymbol{I}$ and $\boldsymbol{X}\boldsymbol{Y}^T=\boldsymbol{0}$ due to the orthonormality property of $\boldsymbol{V}$. That finally proves Theorem B.2. □

B.2 Generic Weighting Matrix for Dynamically Consistent Pseudoinverses

The proof for dynamic consistency of the pseudoinverse with the weighting matrix (4.33) is provided in the following. The Theorems B.1 and B.2 from Sect. B.1 are employed.

Theorem B.3 *A redundant n-DOF manipulator with symmetric and positive definite inertia matrix* $\boldsymbol{M}\in\mathbb{R}^{n\times n}$ *and non-singular, primary task Jacobian matrix* $\boldsymbol{J}\in\mathbb{R}^{m\times n}$ *(for* $m<n$*) with singular value decomposition* (B.1)–(B.3) *is considered. Using the weighting matrix*

$$\boldsymbol{W}=\boldsymbol{X}^T\boldsymbol{X}\boldsymbol{B}_X+\boldsymbol{B}_Y\boldsymbol{Y}^T\boldsymbol{Y}\boldsymbol{M}\ \ \textit{with}\ \ \operatorname{rank}(\boldsymbol{B}_X)\geq m\ \wedge\ \operatorname{rank}(\boldsymbol{B}_Y)\geq n-m$$

in the pseudoinversion (4.15) of $\boldsymbol{J}$ *leads to a dynamically consistent solution (cf. (4.23)), thus the following identity must hold for the null space projector* $\boldsymbol{N}\in\mathbb{R}^{n\times n}$ *in the null space of* $\boldsymbol{J}$*:*

$$\boldsymbol{J}\boldsymbol{M}^{-1}\boldsymbol{N}=\boldsymbol{0}\,.$$

Proof Equation (B.8) is used in this proof, i.e. $\boldsymbol{N}=\boldsymbol{W}^T\boldsymbol{Y}^T(\boldsymbol{Y}\boldsymbol{W}^T\boldsymbol{Y}^T)^{-1}\boldsymbol{Y}$. Due to $\boldsymbol{X}\boldsymbol{Y}^T=\boldsymbol{0}$, the simplification

$$\begin{aligned}\boldsymbol{W}^T\boldsymbol{Y}^T&=(\boldsymbol{B}_X^T\boldsymbol{X}^T\boldsymbol{X}+\boldsymbol{M}\boldsymbol{Y}^T\boldsymbol{Y}\boldsymbol{B}_Y^T)\boldsymbol{Y}^T\\&=\boldsymbol{M}\boldsymbol{Y}^T\boldsymbol{Y}\boldsymbol{B}_Y^T\boldsymbol{Y}^T\end{aligned} \tag{B.12}$$

can be made. Dynamic consistency can then be shown:

$$\begin{aligned} JM^{-1}N &= JM^{-1}MY^TYB_Y^TY^T(YW^TY^T)^{-1}Y \\ &= USV^TY^TYB_Y^TY^T(YW^TY^T)^{-1}Y \\ &= U\,(\Sigma\,,\mathbf{0})\begin{pmatrix}X\\Y\end{pmatrix}Y^TYB_Y^TY^T(YW^TY^T)^{-1}Y \\ &= U\,(\Sigma\,,\mathbf{0})\begin{pmatrix}\mathbf{0}\\I\end{pmatrix}YB_Y^TY^T(YW^TY^T)^{-1}Y \\ &= \mathbf{0}\,. \end{aligned}$$

□

If (B.8) is utilized for the projector computation instead of (B.4) or (B.7), then the restrictions on the rank of B_X can even be loosened. Since (B.8) does not require W^{-1}, only fulfilling $\text{rank}(B_Y) \geq n - m$ is necessary. Again, this condition is necessary but not sufficient.

The conclusions obtained in the context of this analysis allow a better understanding of the null space projectors in general. Actually, knowledge of the inertia matrix in the pseudoinversion is only required applied to the null space as it can be seen in the general formulation of the weighting matrix if $B_X = \mathbf{0}$. Note that the presented proofs of Theorems B.1 and B.2 do not cover the loosened rank conditions on B_X since inversion of W is assumed there. However, it is evident from (B.8) that the inversion is feasible as long as $\text{rank}(YW^TY^T) = \text{rank}(Y)$. Due to (B.12), this inversion simplifies to $(YMY^TYB_Y^TY^T)^{-1}$, where B_X does not appear at all.

B.3 Dynamic Consistency for an Arbitrary, Invertible Weighting Matrix *W*

In this section, dynamic consistency of the projection (4.38) according to Definition 4.2 is shown.

$$\begin{aligned} JM^{-1}N &= JM^{-1}M(I - J^{W+}J)M^{-1} \\ &= (J - \underbrace{JW^{-1}J^T(JW^{-1}J^T)^{-1}}_{= I \text{ for } \text{rank}(W) = n}J)M^{-1} \\ &= \mathbf{0}\,. \end{aligned} \tag{B.13}$$

Moreover, one can show that any weighting matrix W fulfilling (4.33) can be used in the acceleration-based approach (4.38) and still yields the classical dynamically consistent null space projector [Kha87].

$$\overbrace{\boldsymbol{M}\left(\boldsymbol{I}-\boldsymbol{J}^{\boldsymbol{M}+}\boldsymbol{J}\right)\boldsymbol{M}^{-1}}^{\text{(4.38) with } \boldsymbol{W} \text{ from (4.33)}} = \boldsymbol{I}-\boldsymbol{J}^{T}\left(\boldsymbol{J}\boldsymbol{M}^{-1}\boldsymbol{J}^{T}\right)^{-1}\boldsymbol{J}\boldsymbol{M}^{-1} = \underbrace{\boldsymbol{I}-\boldsymbol{J}^{T}(\boldsymbol{J}^{\boldsymbol{M}+})^{T}}_{\text{[Kha87]}}. \tag{B.14}$$

Appendix C
Proofs for the Stability Analysis

The following sections present a proof for the validity of $\boldsymbol{Z}_i$ in Sect. C.1, the invertibility of the extended Jacobian matrix $\bar{\boldsymbol{J}}$ in Sect. C.2, and the block-diagonal structure of the inertia matrix $\boldsymbol{\Lambda}$ in Sect. C.3. For the sake of simplicity, the dependencies on $\boldsymbol{q}$ are omitted in the notations.

C.1 Derivation of $\boldsymbol{Z}_i(\boldsymbol{q})$

In this section the validity of (5.51) is shown via the following Lemma:

Lemma C.1 *If $\bar{\boldsymbol{J}}_i$ and $\boldsymbol{Z}_i$ for all levels ($1 \leq i \leq r$) are chosen following (5.44) and (5.51)*

$$\bar{\boldsymbol{J}}_i = \left(\boldsymbol{Z}_i \boldsymbol{M} \boldsymbol{Z}_i^T\right)^{-1} \boldsymbol{Z}_i \boldsymbol{M} = (\boldsymbol{Z}_i^{\boldsymbol{M}^{-1}+})^T,$$

$$\boldsymbol{Z}_i = \begin{cases} (\boldsymbol{J}_1^{\boldsymbol{M}+})^T & \text{if } i = 1 \\ \boldsymbol{J}_i \boldsymbol{M}^{-1} \boldsymbol{N}_i & \text{if } 2 \leq i < r \\ \boldsymbol{Y}_{r-1} & \text{if } i = r \end{cases},$$

then the identity

$$\bar{\boldsymbol{J}}_i^T \boldsymbol{Z}_i \boldsymbol{J}_i^T = \boldsymbol{N}_i \boldsymbol{J}_i^T . \tag{C.1}$$

holds, where the dynamically consistent null space projector $\boldsymbol{N}_i$ is defined in (5.45) for $2 \leq i \leq r$, and $\boldsymbol{N}_1 = \boldsymbol{I}$ since no restriction is set on the main task.

Proof Three cases have to be considered to cover all levels required in Lemma C.1. That contains the main task $i = 1$, the range $2 \leq i < r$, and the bottom priority level $i = r$.

A. Dietrich, *Whole-Body Impedance Control of Wheeled Humanoid Robots*,
Springer Tracts in Advanced Robotics 116, DOI 10.1007/978-3-319-40557-5

Case $i = 1$ ***with*** $\boldsymbol{Z}_1 = (\boldsymbol{J}_1^{M+})^T$***:***

$$
\begin{aligned}
\bar{\boldsymbol{J}}_1^T \boldsymbol{Z}_1 \boldsymbol{J}_1^T &= \boldsymbol{M}\boldsymbol{J}_1^{M+}\left((\boldsymbol{J}_1^{M+})^T \boldsymbol{M}\boldsymbol{J}_1^{M+}\right)^{-1}(\boldsymbol{J}_1^{M+})^T \boldsymbol{J}_i^T \\
&= \boldsymbol{J}_1^T\left(\boldsymbol{J}_1\boldsymbol{M}^{-1}\boldsymbol{J}_1^T\right)^{-1}\left(\left(\boldsymbol{J}_1\boldsymbol{M}^{-1}\boldsymbol{J}_1^T\right)^{-1}\boldsymbol{J}_1\boldsymbol{M}^{-1}\boldsymbol{J}_1^T\left(\boldsymbol{J}_1\boldsymbol{M}^{-1}\boldsymbol{J}_1^T\right)^{-1}\right)^{-1} \\
&= \boldsymbol{N}_1\boldsymbol{J}_1^T\,.
\end{aligned}
\tag{C.2}
$$

Case $2 \le i < r$ ***with*** $\boldsymbol{Z}_i = \boldsymbol{J}_i\boldsymbol{M}^{-1}\boldsymbol{N}_i$***:***

$$
\begin{aligned}
\bar{\boldsymbol{J}}_i^T \boldsymbol{Z}_i \boldsymbol{J}_i^T &= \boldsymbol{M}\boldsymbol{Z}_i^T\left(\boldsymbol{Z}_i\boldsymbol{M}\boldsymbol{Z}_i^T\right)^{-1}\boldsymbol{Z}_i\boldsymbol{J}_i^T \\
&= \boldsymbol{M}\boldsymbol{N}_i^T\boldsymbol{M}^{-1}\boldsymbol{J}_i^T\left(\boldsymbol{J}_i\boldsymbol{M}^{-1}\boldsymbol{N}_i\boldsymbol{M}\boldsymbol{N}_i^T\boldsymbol{M}^{-1}\boldsymbol{J}_i^T\right)^{-1}\boldsymbol{J}_i\boldsymbol{M}^{-1}\boldsymbol{N}_i\boldsymbol{J}_i^T \\
&= \boldsymbol{M}\boldsymbol{Y}_{i-1}^T\left(\boldsymbol{Y}_{i-1}\boldsymbol{M}\boldsymbol{Y}_{i-1}^T\right)^{-1}\boldsymbol{Y}_{i-1}\boldsymbol{J}_i^T \\
&\quad \left(\boldsymbol{J}_i\boldsymbol{Y}_{i-1}^T\left(\boldsymbol{Y}_{i-1}\boldsymbol{M}\boldsymbol{Y}_{i-1}^T\right)^{-1}\boldsymbol{Y}_{i-1}\boldsymbol{M}\boldsymbol{Y}_{i-1}^T\left(\boldsymbol{Y}_{i-1}\boldsymbol{M}\boldsymbol{Y}_{i-1}^T\right)^{-1}\boldsymbol{Y}_{i-1}\boldsymbol{J}_i^T\right)^{-1} \\
&\quad \boldsymbol{J}_i\boldsymbol{Y}_{i-1}^T\left(\boldsymbol{Y}_{i-1}\boldsymbol{M}\boldsymbol{Y}_{i-1}^T\right)^{-1}\boldsymbol{Y}_{i-1}\boldsymbol{J}_i^T \\
&= \boldsymbol{M}\boldsymbol{Y}_{i-1}^T\left(\boldsymbol{Y}_{i-1}\boldsymbol{M}\boldsymbol{Y}_{i-1}^T\right)^{-1}\boldsymbol{Y}_{i-1}\boldsymbol{J}_i^T\left(\boldsymbol{J}_i\boldsymbol{Y}_{i-1}^T\left(\boldsymbol{Y}_{i-1}\boldsymbol{M}\boldsymbol{Y}_{i-1}^T\right)^{-1}\boldsymbol{Y}_{i-1}\boldsymbol{J}_i^T\right)^{-1} \\
&\quad \boldsymbol{J}_i\boldsymbol{Y}_{i-1}^T\left(\boldsymbol{Y}_{i-1}\boldsymbol{M}\boldsymbol{Y}_{i-1}^T\right)^{-1}\boldsymbol{Y}_{i-1}\boldsymbol{J}_i^T \\
&= \boldsymbol{M}\boldsymbol{Y}_{i-1}^T\left(\boldsymbol{Y}_{i-1}\boldsymbol{M}\boldsymbol{Y}_{i-1}^T\right)^{-1}\boldsymbol{Y}_{i-1}\boldsymbol{J}_i^T \\
&= \boldsymbol{N}_i\boldsymbol{J}_i^T\,.
\end{aligned}
\tag{C.3}
$$

Case $i = r$ ***with*** $\boldsymbol{Z}_r = \boldsymbol{Y}_{r-1}$***:***

$$
\begin{aligned}
\bar{\boldsymbol{J}}_r^T \boldsymbol{Z}_r \boldsymbol{J}_r^T &= \boldsymbol{M}\boldsymbol{Z}_r^T\left(\boldsymbol{Z}_r\boldsymbol{M}\boldsymbol{Z}_r^T\right)^{-1}\boldsymbol{Z}_r\boldsymbol{J}_r^T \\
&= \boldsymbol{M}\boldsymbol{Y}_{r-1}^T\left(\boldsymbol{Y}_{r-1}\boldsymbol{M}\boldsymbol{Y}_{r-1}^T\right)^{-1}\boldsymbol{Y}_{r-1}\boldsymbol{J}_r^T \\
&= \boldsymbol{N}_r\boldsymbol{J}_r^T\,.
\end{aligned}
\tag{C.4}
$$

□

C.2 Extended Jacobian Matrix $\bar{\boldsymbol{J}}$ and its Inverse

In this section, the invertibility of the extended Jacobian matrix $\bar{\boldsymbol{J}} \in \mathbb{R}^{n\times n}$ is shown through the existence of its inverse $\bar{\boldsymbol{J}}^{-1}$. According to Chap. 5, the extended Jacobian matrix has the form

$$\bar{J} = \begin{pmatrix} \bar{J}_1 \\ \bar{J}_2 \\ \vdots \\ \bar{J}_r \end{pmatrix} = \begin{pmatrix} J_1 \\ \left(Z_2 M Z_2^T\right)^{-1} Z_2 M \\ \vdots \\ \left(Z_r M Z_r^T\right)^{-1} Z_r M \end{pmatrix} \tag{C.5}$$

with the following properties:

- The matrices $J_1 \dots J_r$ have full row rank. They are used in the computation of the full-row-rank matrices $Z_i \ \forall i,\ 1 \le i \le r$.
- The matrices $\bar{J}_1 \dots \bar{J}_r$ and consequently the augmented Jacobian matrices J_i^{aug} for $i = 1 \dots r-1$ have full row rank. No algorithmic singularities are encountered.
- The inertia matrix M is symmetric and positive definite.
- The row vectors of Z_i span the (right) null space of all higher-level Jacobian matrices $J_1 \dots J_{i-1}$ and $\bar{J}_1 \dots \bar{J}_{i-1}$, thus

$$J_i Z_j^T = \mathbf{0}\,, \tag{C.6}$$

$$\bar{J}_i Z_j^T = \mathbf{0}\,, \tag{C.7}$$

for $i < j$. The argumentation to (C.15) in Sect. C.3 will implicitly comprise the proof of (C.7).

Lemma C.2 *If the extended Jacobian matrix* (C.5) *has the properties listed above, then it is non-singular and invertible due to the existence of its inverse* $\bar{J}^{-1}$.

Proof Suppose that $\bar{J}^{-1}$ has the form

$$\bar{J}^{-1} = \left(C_1\ C_2 \cdots C_r\right),$$

where $C_i \in \mathbb{R}^{n \times m_i}$ for $i = 1 \dots r$ is not known so far. The identity

$$\bar{J}\bar{J}^{-1} = \begin{pmatrix} J_1 C_1 & J_1 C_2 & \cdots & J_1 C_r \\ \left(Z_2 M Z_2^T\right)^{-1} Z_2 M C_1 & \left(Z_2 M Z_2^T\right)^{-1} Z_2 M C_2 & & \left(Z_2 M Z_2^T\right)^{-1} Z_2 M C_r \\ \vdots & & \ddots & \vdots \\ \left(Z_r M Z_r^T\right)^{-1} Z_r M C_1 & \left(Z_r M Z_r^T\right)^{-1} Z_r M C_2 & \cdots & \left(Z_r M Z_r^T\right)^{-1} Z_r M C_r \end{pmatrix} = I$$

must hold for an invertible matrix $\bar{J}$. The matrix $C_1 = J_1^{M+}$ fulfills the first column requirements

$$I = J_1 J_1^{M+}, \tag{C.8}$$

$$\mathbf{0} = \left(Z_i M Z_i^T\right)^{-1} Z_i M J_1^{M+} \tag{C.9}$$

$$= \left(Z_i M Z_i^T\right)^{-1} \underbrace{Z_i J_1^T}_{=\mathbf{0}} \left(J_1 M^{-1} J_1^T\right)^{-1}, \tag{C.10}$$

for $1 < i \leq r$. The matrices $\boldsymbol{C}_j = \boldsymbol{Z}_j^T \; \forall j, \; 1 < j \leq r$ fulfill the remaining first row requirements

$$\mathbf{0} = \boldsymbol{J}_1 \boldsymbol{Z}_j^T . \tag{C.11}$$

What remains is to show that

$$\boldsymbol{I} = \left(\boldsymbol{Z}_i \boldsymbol{M} \boldsymbol{Z}_i^T\right)^{-1} \boldsymbol{Z}_i \boldsymbol{M} \boldsymbol{Z}_i^T , \tag{C.12}$$

$$\mathbf{0} = \left(\boldsymbol{Z}_i \boldsymbol{M} \boldsymbol{Z}_i^T\right)^{-1} \underbrace{\boldsymbol{Z}_i \boldsymbol{M} \boldsymbol{Z}_j^T}_{=\mathbf{0}} \tag{C.13}$$

is valid for the remaining elements with $i \neq j$. While (C.12) is obvious, proving the identity (C.13) is rather complex. The annihilation $\boldsymbol{Z}_i \boldsymbol{M} \boldsymbol{Z}_j^T = \mathbf{0}$ will be shown in Appendix C.3 in the context of inertia decoupling (C.15). □

C.3 Decoupling in the Inertia Matrix $\boldsymbol{\Lambda}$

The proof of the analogous statements (5.42), (5.43)

$$\bar{\boldsymbol{J}}_i \boldsymbol{M}^{-1} \bar{\boldsymbol{J}}_j^T = \mathbf{0} , \tag{C.14}$$

$$\boldsymbol{Z}_i \boldsymbol{M} \boldsymbol{Z}_j^T = \mathbf{0} \tag{C.15}$$

for $i \neq j$ is provided in the following. Their physical interpretation is the decoupling of the inertias $\boldsymbol{\Lambda}_i$ on different priority levels. Thus, the inertia matrix is of block-diagonal structure. According to (5.51), the null space base matrices $\boldsymbol{Z}_i$ are

$$\boldsymbol{Z}_i = \begin{cases} (\boldsymbol{J}_1^{M+})^T & \text{for } i = 1 \\ \boldsymbol{J}_i \boldsymbol{M}^{-1} \boldsymbol{N}_i & \text{for } 1 < i < r \\ \boldsymbol{Y}_{r-1} & \text{for } i = r \end{cases} \tag{C.16}$$

For all null space levels, the standard dynamically consistent projector [Kha87] has the form

$$\boldsymbol{N}_i = \boldsymbol{M} \boldsymbol{Y}_{i-1}^T \left(\boldsymbol{Y}_{i-1} \boldsymbol{M} \boldsymbol{Y}_{i-1}^T\right)^{-1} \boldsymbol{Y}_{i-1} \tag{C.17}$$

as shown in Theorem B.2. Now the condition (C.15) is verified for all possible values of i and j.

Case $i = 1, \; 1 < j < r$ ***and the transposed element:***

$$\left(\boldsymbol{J}_1 \boldsymbol{M}^{-1} \boldsymbol{J}_1^T\right)^{-1} \underbrace{\boldsymbol{J}_1 \boldsymbol{Y}_{j-1}^T}_{=\mathbf{0}} \left(\boldsymbol{Y}_{j-1} \boldsymbol{M} \boldsymbol{Y}_{j-1}^T\right)^{-1} \boldsymbol{Y}_{j-1} \boldsymbol{J}_j^T = \mathbf{0} .$$

Case $i = 1,\ j = r$ ***and the transposed element:***

$$\left(\boldsymbol{J}_1\boldsymbol{M}^{-1}\boldsymbol{J}_1^T\right)^{-1}\underbrace{\boldsymbol{J}_1\boldsymbol{Y}_{r-1}^T}_{=\boldsymbol{0}} = \boldsymbol{0}\,.$$

Case $1 < i < r,\ 1 < j < r$ ***with*** $i \neq j$***:***

$$\boldsymbol{J}_i\boldsymbol{M}^{-1}\boldsymbol{M}\boldsymbol{Y}_{i-1}^T\left(\boldsymbol{Y}_{i-1}\boldsymbol{M}\boldsymbol{Y}_{i-1}^T\right)^{-1}\boldsymbol{Y}_{i-1}\boldsymbol{M}\boldsymbol{Y}_{j-1}^T\left(\boldsymbol{Y}_{j-1}\boldsymbol{M}\boldsymbol{Y}_{j-1}^T\right)^{-1}\boldsymbol{Y}_{j-1}\boldsymbol{J}_j^T = \boldsymbol{0}\,,$$

which can then be rephrased and simplified to

$$\boldsymbol{J}_i\boldsymbol{M}^{-1}\boldsymbol{N}_{\max(i,j)}\boldsymbol{J}_j^T = \boldsymbol{0}\,. \tag{C.18}$$

Note that $\boldsymbol{N}_{\max(i,j)} = \boldsymbol{N}_i\boldsymbol{N}_j$ can be concluded from the idempotence of the null space projectors. Based on that, there are two different cases to be considered: $i < j$ and $i > j$. For $i < j$, (C.18) becomes

$$\underbrace{\boldsymbol{J}_i\boldsymbol{M}^{-1}\boldsymbol{N}_j}_{=\boldsymbol{0}}\boldsymbol{J}_j^T = \boldsymbol{0} \tag{C.19}$$

and fulfills the criterion due to the dynamic consistency of the null space projector $\boldsymbol{N}_j$, cf. Definition 4.2. For $i > j$, (C.18) becomes

$$\boldsymbol{J}_i\boldsymbol{M}^{-1}\underbrace{\boldsymbol{N}_i\boldsymbol{J}_j^T}_{=\boldsymbol{0}} = \boldsymbol{0}\,, \tag{C.20}$$

which is a direct consequence of $\boldsymbol{Y}_{i-1}\boldsymbol{J}_j^T = \boldsymbol{0}$ for $i > j$ or the fact that $\boldsymbol{J}_j$ was actually used to determine its null space $\boldsymbol{Y}_{i-1}$ and the projector $\boldsymbol{N}_i$, respectively.

Case $i = r,\ 1 < j < r$ ***and the transposed element:***

$$\boldsymbol{Y}_{r-1}\boldsymbol{M}\boldsymbol{Y}_{j-1}^T\left(\boldsymbol{Y}_{j-1}\boldsymbol{M}\boldsymbol{Y}_{j-1}^T\right)^{-1}\boldsymbol{Y}_{j-1}\boldsymbol{J}_j^T = \boldsymbol{0} \tag{C.21}$$

$$\boldsymbol{Y}_{r-1}\boldsymbol{N}_j\boldsymbol{J}_j^T = \boldsymbol{0}\,. \tag{C.22}$$

The multiplication by the identity $(\boldsymbol{Y}_{r-1}\boldsymbol{M}\boldsymbol{Y}_{r-1}^T)(\boldsymbol{Y}_{r-1}\boldsymbol{M}\boldsymbol{Y}_{r-1}^T)^{-1} = \boldsymbol{I}$ from the left yields

$$\boldsymbol{Y}_{r-1}\boldsymbol{M}\boldsymbol{Y}_{r-1}^T(\boldsymbol{Y}_{r-1}\boldsymbol{M}\boldsymbol{Y}_{r-1}^T)^{-1}\boldsymbol{Y}_{r-1}\boldsymbol{N}_j\boldsymbol{J}_j^T = \boldsymbol{0} \tag{C.23}$$

$$\boldsymbol{Y}_{r-1}\boldsymbol{N}_r\boldsymbol{N}_j\boldsymbol{J}_j^T = \boldsymbol{0} \tag{C.24}$$

$$\boldsymbol{Y}_{r-1}\underbrace{\boldsymbol{N}_r\boldsymbol{J}_j^T}_{=\boldsymbol{0}} = \boldsymbol{0}\,. \tag{C.25}$$

The cancellation of $\boldsymbol{N}_j$ in (C.24) is due to the idempotence of the null space projector, cf. (C.18). The annihilation in (C.25) can be justified in a similar way as the one in (C.20).

Appendix D
Stability Definitions

The following definitions are extracted from [SJK97, vdS00, OKN08, ODAS15] and explain the notions of *strict output passivity* and *conditional stability*.

D.1 Passivity

Definition D.1 *A system $\dot{z} = f(z, u)$ with input u and output y is said to be strictly output passive if there exists a non-negative function $S(z)$ and an $\epsilon > 0$ such that*

$$S(z(t)) - S(z(0)) \leq \int_0^t (y(s)^T u(s) - \epsilon \, \|y(s)\|^2) ds. \tag{D.1}$$

holds [vdS00] for all $t > 0$.

Definition D.2 *A system $\dot{z} = f(z, u)$ with input u and output y is said to be strictly output passive conditionally to $\mathcal{A} \subset \mathbb{R}^n$, if it is strictly output passive for any initial condition $z_0 = z(0) \in \mathcal{A}$.*

D.2 Conditional Stability

A time-invariant system with state vector $z \in \mathbb{R}^n$ has the form

$$\dot{z} = f(z) \,. \tag{D.2}$$

The state z_s is a stationary point of (D.2) so that $f(z_s) = 0$. Assume that there exists a solution $z(t)$ for (D.2) for an initial state $z_0 = z(0)$ for all times $t > 0$. For conditional stability all requirements of the stability definitions must only hold

A. Dietrich, *Whole-Body Impedance Control of Wheeled Humanoid Robots*,
Springer Tracts in Advanced Robotics 116, DOI 10.1007/978-3-319-40557-5

for those initial conditions which lie in a particular set $\mathcal{A} \subset \mathbb{R}^n$. Thus conditional stability is weaker than the usual Lyapunov stability.

Definition D.3 *A stationary point* z_s *of* (D.2) *is said to be stable conditionally to the set* $\mathcal{A} \subset \mathbb{R}^n$, *if* $z_s \in \mathcal{A}$ *and for each* $\epsilon > 0$ *there exists a* $\delta(\epsilon) > 0$ *such that the following implication holds for any initial condition* $z_0 \in \mathcal{A}$:

$$\|z_0 - z_s\| < \delta(\epsilon) \implies \|z(t) - z_s\| < \epsilon\ , \quad \forall t \geq 0\ . \tag{D.3}$$

Definition D.4 *A stationary point* z_s *of* (D.2) *is said to be attractive conditionally to* $\mathcal{A} \subset \mathbb{R}^n$, *if* $z_s \in \mathcal{A}$ *and there exists a* $\eta(z_s) > 0$ *such that the following implication holds for any initial condition* $z_0 \in \mathcal{A}$:

$$\|z_0 - z_s\| < \eta(z_s) \implies \lim_{t\to\infty} z(t) = z_s\ . \tag{D.4}$$

Definition D.5 *A stationary point* z_s *of* (D.2) *is said to be asymptotically stable conditionally to* $\mathcal{A} \subset \mathbb{R}^n$, *if it is both stable and attractive conditionally to* $\mathcal{A}$.

Definition D.6 *A stationary point* z_s *of* (D.2) *is said to be globally asymptotically stable conditionally to* $\mathcal{A} \subset \mathbb{R}^n$, *if it is asymptotically stable conditionally to* $\mathcal{A}$ *with* $\eta(z_s) = +\infty$.

References

[AH09] Samuel K. Au and Hugh M. Herr. Powered Ankle-Foot Prosthesis. *IEEE Robotics & Automation Magazine* 15(3):52–59, September 2009

[AIC09] Gianluca Antonelli, Giovanni Indiveri, and Stefano Chiaverini. Prioritized Closed-Loop Inverse Kinematic Algorithms for Redundant Robotic Systems with Velocity Saturations. In *Proc. of the 2009 IEEE/RSJ International Conference on Intelligent Robots and Systems*, pages 5892–5897, October 2009

[Ant09] Gianluca Antonelli. Stability Analysis for Prioritized Closed-Loop Inverse Kinematic Algorithms for Redundant Robotic Systems. *IEEE Transactions on Robotics*, 25(5):985–994, October 2009

[ARS+06] T. Asfour, K. Regenstein, J. Schröder, A. Bierbaum, N. Vahrenkamp, and R. Dillmann. ARMAR-III: An Integrated Humanoid Platform for Sensory-Motor Control. In *Proc. of the 6th IEEE-RAS International Conference on Humanoid Robots*, pages 169–175, December 2006

[ASEG+08] Alin Albu-Schäffer, Oliver Eiberger, Markus Grebenstein, Sami Haddadin, Christian Ott, Thomas Wimböck, Sebastian Wolf and Gerd Hirzinger. Soft Robotics: From Torque Feedback-Controlled Lightweight Robots to Instrinsically Compliant Systems. *IEEE Robotics & Automation Magazine*, 15(3):20–30, September 2008

[Asi42] Isaac Asimov. *Runaround.* Robot Series. Street & Smith, March 1942

[ASOFH03] Alin Albu-Schäffer, Christian Ott, Udo Frese, and Gerd Hirzinger. Cartesian Impedance Control of Redundant Robots: Recent Results with the DLR-Light-Weight-Arms. In *Proc. of the 2003 IEEE International Conference on Robotics and Automation*, pages 3704–3709, Sept. 2003

[ASOH04] Alin Albu-Schäffer, Christian Ott, and Gerd Hirzinger. A Passivity Based Cartesian Impedance Controller for Flexible Joint Robots - Part II: Full State Feedback, Impedance Design and Experiments. In *Proc. of the 2004 IEEE International Conference on Robotics and Automation*, pages 2666–2672, April 2004

[ASOH07] Alin Albu-Schäffer, Christian Ott, and Gerd Hirzinger. A Unified Passivity-based Control Framework for Position, Torque and Impedance Control of Flexible Joint Robots. *International Journal of Robotics Research*, 27(1):23–39, January 2007

[Bai85] John Baillieul. Kinematic programming alternatives for redundant manipulators. In *Proc. of the 1985 IEEE International Conference on Robotics and Automation*, pages 722–728, March 1985

[BB98] P. Baerlocher and R. Boulic. Task-Priority Formulations for the Kinematic Control of Highly Redundant Articulated Structures. In *Proc. of the 1998 IEEE/RSJ International Conference on Intelligent Robots and Systems*, pages 323–329, October 1998

A. Dietrich, *Whole-Body Impedance Control of Wheeled Humanoid Robots*,
Springer Tracts in Advanced Robotics 116, DOI 10.1007/978-3-319-40557-5

[BB04] Paolo Baerlocher and Ronan Boulic. An Inverse Kinematic Architecture Enforcing an Arbitrary Number of Strict Priority Levels. *The Visual Computer*, 20(6):402–417, August 2004

[BBW+11] Berthold Bäuml, Oliver Birbach, Thomas Wimböck, Udo Frese, Alexander Dietrich, and Gerd Hirzinger. Catching Flying Balls with a Mobile Humanoid: System Overview and Design Considerations. In *Proc. of the 11th IEEE-RAS International Conference on Humanoid Robots*, pages 513–520, October 2011

[BGLH01] Jörg Butterfass, Markus Grebenstein, H. Liu, and Gerd Hirzinger. DLR-Hand II: Next Generation of a Dextrous Robot Hand. In *Proc. of the 2001 IEEE International Conference on Robotics and Automation*, pages 109–114, May 2001

[BHB84] John Baillieul, John M. Hollerbach, and Roger Brockett. Programming and control of kinematically redundant manipulators. In *Proc. of the 23rd IEEE Conference on Decision and Control*, pages 768–774, December 1984

[BHG10] Matthias Behnisch, Robert Haschke, and Michael Gienger. Task Space Motion Planning Using Reactive Control. In *Proc. of the 2010 IEEE/RSJ International Conference on Intelligent Robots and Systems*, pages 5934–5940, October 2010

[BK02] Oliver Brock, and Oussama Khatib. Elastic Strips: A Framework for Motion Generation in Human Environments. *International Journal of Robotics Research* 21(12):1031–1052, December 2002

[BKV02] Oliver Brock, Oussama Khatib, and Sriram Viji. Task-Consistent Obstacle Avoidance and Motion Behavior for Mobile Manipulation. In *Proc. of the 2002 IEEE International Conference on Robotics and Automation*, pages 388–393, May 2002

[BOW+07] Christoph Borst, Christian Ott, Thomas Wimböck, Bernhard Brunner, Franziska Zacharias, Berthold Bäuml, Ulrich Hillenbrand, Sami Haddadin, Alin Albu-Schäffer, and Gerd Hirzinger. A humanoid upper-body system for two-handed manipulation. In *Proc. of the 2007 IEEE International Conference on Robotics and Automation*, pages 2766–2767, April 2007

[BRJ+11] Jonathan Bohren, Radu Bogdan Rusu, E. Gil Jones, Eitan Marder-Eppstein, Caroline Pantofaru, Melonee Wise, Lorenz Mösenlechner, Wim Meeussen, and Stefan Holzer. Towards Autonomous Robotic Butlers: Lessons Learned with the PR2. In *Proc. of the 2011 IEEE International Conference on Robotics and Automation*, pages 5568–5575, May 2011

[BSW+11] Berthold Bäuml, Florian Schmidt, Thomas Wimböck, Oliver Birbach, Alexander Dietrich, Matthias Fuchs, Werner Friedl, Udo Frese, Christoph Borst, Markus Grebenstein, Oliver Eiberger, and Gerd Hirzinger. Catching Flying Balls and Preparing Coffee: Humanoid Rollin' Justin Performs Dynamic and Sensitive Tasks. In *Proc. of the 2011 IEEE International Conference on Robotics and Automation*, pages 3443–3444, May 2011

[Bus14] Kristin Bussmann. Whole-Body Stability for a Humanoid Robot: Analysis, Control Design, and Experimental Evaluation. Master's thesis, Technische Universität München, 2014

[BWS+09] Christoph Borst, Thomas Wimböck, Florian Schmidt, Matthias Fuchs, Bernhard Brunner, Franziska Zacharias, Paolo Robuffo Giordano, Rainer Konietschke, Wolfgang Sepp, Stefan Fuchs, Christian Rink, Alin Albu-Schäffer, and Gerd Hirzinger. Rollin' Justin - Mobile Platform with Variable Base. In *Proc. of the 2009 IEEE International Conference on Robotics and Automation*, pages 1597–1598, May 2009

[CBDN96] Guy Campion, Georges Bastin, and Brigitte D'Andréa-Novel. Structural Properties and Classification of Kinematic and Dynamic Models of Wheeled Mobile Robots. *IEEE Transactions on Robotics and Automation*, 12(1):47–62, February 1996

[CD95] Tan Fung Chan and Rajiv V. Dubey. A Weighted Least-Norm Solution Based Scheme for Avoiding Joint Limits for Redundant Joint Manipulators. *IEEE Transactions on Robotics and Automation* 11(2):286–292, April 1995

[CDG14] Maxime Chalon, Alexander Dietrich, and Markus Grebenstein. The Thumb of the Anthropomorphic Awiwi Hand: From Concept to Evaluation. *International Journal of Humanoid Robotics*, 11(3), 2014

[CH89] Ed Colgate and Neville Hogan. An analysis of contact instability in terms of passive physical equivalents. In *Proc. of the 1989 IEEE International Conference on Robotics and Automation*, pages 404–409, May 1989

[Chi97] Stefano Chiaverini. Singularity-Robust Task-Priority Redundancy Resolution for Real-Time Kinematic Control of Robot Manipulators. *IEEE Transactions on Robotics and Automation*, 13(3):398–410, June 1997

[CK95] K.-S. Chang and Oussama Khatib. Manipulator Control at Kinematic Singularities: A Dynamically Consistent Strategy. In *Proc. of the 1995 IEEE/RSJ International Conference on Intelligent Robots and Systems*, pages 84–88, August 1995

[CPHV08] Christian P. Connette, Andreas Pott, Martin Hägele, and Alexander Verl. Control of an Pseudo-omnidirectional, Non-holonomic, Mobile Robot based on an ICM Representation in Spherical Coordinates. In *Proc. of the 47th IEEE Conference on Decision and Control*, pages 4976–4983, December 2008

[CPHV09] Christian P. Connette, Christopher Parlitz, Martin Hägele, and Alexander Verl. Singularity Avoidance for Over-Actuated, Pseudo-Omnidirectional, Wheeled Mobile Robots. In *Proc. of the 2009 IEEE International Conference on Robotics and Automation*, pages 4124–4130, May 2009

[Cra89] John Craig. *Introduction to Robotics: Mechanics and Control*. Addison-Wesley, 1989

[CW93] Yu-Che Chen and Ian D. Walker. A Consistent Null-Space Based Approach to Inverse Kinematics of Redundant Robots. In *Proc. of the 1993 IEEE International Conference on Robotics and Automation*, pages 374–381, May 1993

[DASH12] Alexander Dietrich, Alin Albu-Schäffer, and Gerd Hirzinger. On Continuous Null Space Projections for Torque-Based, Hierarchical, Multi-Objective Manipulation. In *Proc. of the 2012 IEEE International Conference on Robotics and Automation*, pages 2978–2985, May 2012

[DBOAS14] Alexander Dietrich, Kristin Bussmann, Christian Ott, and Alin Albu-Schäffer. Ganzkörperimpedanz für mobile Roboter. German patent application No. 10 2014 226 936, patented on December 23, 2014

[DBP+16] Alexander Dietrich, Kristin Bussmann, Florian Petit, Paul Kotyczka, Christian Ott, Boris Lohmann and Alin Albu-Schäffer. Whole-body impedance control of wheeled mobile manipulators: Stability analysis and experiments on the humanoid robot Rollin' Justin. *Autonomous Robots* 40(3):505–517, March 2016

[DGTN09] Christian Dornhege, Marc Gissler, Matthias Teschner, and Bernhard Nebel. Integrating Symbolic and Geometric Planning for Mobile Manipulation. In *2009 IEEE International Workshop on Safety, Security & Rescue Robotics*, pages 1–6, November 2009

[DH08] Aaron M. Dollar and Hugh Herr. Lower Extremity Exoskeletons and Active Orthoses: Challenges and State-of-the-Art. *IEEE Transactions on Robotics* 24(1):144–158, February 2008

[DKW+14] Alexander Dietrich, Melanie Kimmel, Thomas Wimböck, Sandra Hirche, and Alin Albu-Schäffer. Workspace Analysis for a Kinematically Coupled Torso of a Torque Controlled Humanoid Robot. In *Proc. of the 2014 IEEE International Conference on Robotics and Automation*, pages 3439–3445, June 2014

[DMA+11] M. Diftler, J. Mehling, M. Abdallah, N. Radford, L. Bridgwater, A. Sanders, R. Askew, D. Linn, J. Yamokoski, F. Permenter, B. Hargrave, R. Platt, R. Savely, and R. Ambrose. Robonaut 2 - The First Humanoid Robot in Space. In *Proc. of the 2011 IEEE International Conference on Robotics and Automation*, pages 2178–2183, May 2011

[DMB93] Keith L. Doty, Claudio Melchiorri, and Claudio Bonivento. A Theory of Generalized Inverses Applied to Robotics. *International Journal of Robotics Research*, 12(1):1–19, February 1993

[DOAS13] Alexander Dietrich, Christian Ott, and Alin Albu-Schäffer. Multi-Objective Compliance Control of Redundant Manipulators: Hierarchy, Control, and Stability. In

Proc. of the 2013 IEEE/RSJ International Conference on Intelligent Robots and Systems, pages 3043–3050, November 2013

[DOAS15] Alexander Dietrich, Christian Ott, and Alin Albu-Schäffer. An overview of null space projections for redundant, torque-controlled robots. *International Journal of Robotics Research* 34(11):1385–1400, September 2015

[DSASO+07] Agostino De Santis, Alin Albu-Schäffer, Christian Ott, Bruno Siciliano, and Gerd Hirzinger. The skeleton algorithm for self-collision avoidance of a humanoid manipulator. In *Proc. of the 2007 IEEE/ASME International Conference on Advanced Intelligent Mechatronics*, September 2007

[DSBDS09] Wilm Decré, Ruben Smits, Herman Bruyninckx, and Joris De Schutter. Extending iTaSC to support inequality constraints and non-instantaneous task specification. In *Proc. of the 2009 IEEE International Conference on Robotics and Automation*, pages 964–971, May 2009

[DSDLR+07] Joris De Schutter, Tinne De Laet, Johan Rutgeerts, Wilm Decré, Ruben Smits, Erwin Artbeliën, Kasper Claes, and Herman Bruyninckx. Constraint-based Task Specification and Estimation for Sensor-Based Robot Systems in the Presence of Geometric Uncertainty. *International Journal of Robotics Research* 26(5):433–455, May 2007

[DW95] Arati Deo, and Ian Walker. Overview of Damped Least-Squares Methods for Inverse Kinematics of Robot Manipulators. *Journal of Intelligent Robotic Systems* 14(1):43–68, September 1995

[DWAS11] Alexander Dietrich, Thomas Wimböck, and Alin Albu-Schäffer. Dynamic Whole-Body Mobile Manipulation with a Torque Controlled Humanoid Robot via Impedance Control Laws. In *Proc. of the 2011 IEEE/RSJ International Conference on Intelligent Robots and Systems*, pages 3199–3206, September 2011

[DWASH11] Alexander Dietrich, Thomas Wimböck, Alin Albu-Schäffer, and Gerd Hirzinger. Singularity Avoidance for Nonholonomic, Omnidirectional Wheeled Mobile Platforms with Variable Footprint. In *Proc. of the 2011 IEEE International Conference on Robotics and Automation*, pages 6136–6142, May 2011

[DWASH12a] Alexander Dietrich, Thomas Wimböck, Alin Albu-Schäffer, and Gerd Hirzinger. Integration of Reactive, Torque-Based Self-Collision Avoidance Into a Task Hierarchy. *IEEE Transactions on Robotics*, 28(6):1278–1293, December 2012

[DWASH12b] Alexander Dietrich, Thomas Wimböck, Alin Albu-Schäffer, and Gerd Hirzinger. Reactive Whole-Body Control: Dynamic Mobile Manipulation Using a Large Number of Actuated Degrees of Freedom. *IEEE Robotics & Automation Magazine* 19(2):20–33, June 2012

[DWT+11] Alexander Dietrich, Thomas Wimböck, Holger Täubig, Alin Albu-Schäffer, and Gerd Hirzinger. Extensions to Reactive Self-Collision Avoidance for Torque and Position Controlled Humanoids. In *Proc. of the 2011 IEEE International Conference on Robotics and Automation*, pages 3455–3462, May 2011

[EC09] Lars-Peter Ellekilde and Henrik I. Christensen. Control of Mobile Manipulator using the Dynamical Systems Approach. In *Proc. of the 2009 IEEE International Conference on Robotics and Automation*, pages 1370–1376, May 2009

[EMW14] Adrien Escande, Nicolas Mansard, and Pierre-Brice Wieber. Hierarchical quadratic programming: Fast online humanoid-robot motion generation. *International Journal of Robotics Research* 33(7):1006–1028, June 2014

[FK97] Roy Featherstone and Oussama Khatib. Load Independence of the Dynamically Consistent Inverse of the Jacobian Matrix. *International Journal of Robotics Research* 16(2):168–170, April 1997

[FW67] Peter L. Falb and William A. Wolovich. Decoupling in the Design and Synthesis of Multivariable Control Systems. *IEEE Transactions on Automatic Control*, AC-12(6):651–659, December 1967

[GEGS12] Martin Grimmer, Mahdy Eslamy, Stefan Gliech, and André Seyfarth. A Comparison of Parallel- and Series Elastic Elements in an actuator for Mimicking Human Ankle

Joint in Walking and Running. In *Proc. of the 2012 IEEE International Conference on Robotics and Automation*, pages 2463–2470, May 2012

[GFASH09] Paolo Robuffo Giordano, Matthias Fuchs, Alin Albu-Schäffer, and Gerd Hirzinger. On the Kinematic Modeling and Control of a Mobile Platform Equipped with Steering Wheels and Movable Legs. In *Proc. of the 2009 IEEE International Conference on Robotics and Automation*, pages 4080–4087, May 2009

[GJK88] Elmer G. Gilbert, Daniel W. Johnson, and S. Sathiya Keerthi. A Fast Procedure for Computing the Distance Between Complex Objects in Three-Dimensional Space. *IEEE Journal of Robotics and Automation*, 4(2):193–203, April 1988

[HASH08] Sami Haddadin, Alin Albu-Schäffer, and Gerd Hirzinger. The Role of the Robot Mass and Velocity in Physical Human-Robot Interaction - Part I: Non-constrained Blunt Impacts. In *Proc. of the 2008 IEEE International Conference on Robotics and Automation*, pages 1331–1338, May 2008

[HASH09] Sami Haddadin, Alin Albu-Schäffer, and Gerd Hirzinger. Requirements for Safe Robots: Measurements, Analysis and New Insights. *International Journal of Robotics Research*, 28(11-12):1507–1527, Nov./Dec. 2009

[HOD10] Ghassan Bin Hammam, David E. Orin, and Behzad Dariush. Whole-Body Humanoid Control from Upper-Body Task Specifications. In *Proc. of the 2010 IEEE International Conference on Robotics and Automation*, pages 3398–3405, May 2010

[Hog85] Neville Hogan. Impedance Control: An Approach to Manipulation: Part I - Theory, Part II - Implementation, Part III - Applications. *Journal of Dynamic Systems, Measurement, and Control*, 107:1–24, March 1985

[HS87] John H. Hollerbach and Ki C. Suh. Redundancy Resolution of Manipulators through Torque Optimization. *IEEE Journal of Robotics and Automation*, RA-3(4):308–316, August 1987

[HSAS+02] Gerd Hirzinger, N. Sporer, Alin Albu-Schäffer, M. Hähnle, R Krenn, A. Pascucci, and M. Schedl. DLR's torque-controlled light weight robot III - are we reaching the technological limits now? In *Proc. of the 2002 IEEE International Conference on Robotics and Automation*, pages 1710–1716, May 2002

[HTSG12] D. F. B. Haeufle, M. D. Taylor, S. Schmitt, and H. Geyer. A clutched parallel elastic actuator concept: towards energy efficient powered legs in prosthetics and robotics. In *Proc. of The Fourth IEEE/RAS/EMBS International Conference on Biomedical Robotics and Biomechatronics*, pages 1614–1619, June 2012

[HV91] Ming Z. Huang and Hareendra Varma. Optimal rate allocation in kinematically redundant manipulators - the dual projection method. In *Proc. of the 1991 IEEE International Conference on Robotics and Automation*, pages 702–707, April 1991

[IKO96] A. Iggidr, B. Kalitine, and R. Outbib. Semidefinite Lyapunov Functions Stability and Stabilization. *Mathematics of Control, Signals, and Systems*, 9(2):95–106, 1996

[IS09] Hiroyasu Iwata and Shigeki Sugano. Design of Human Symbiotic Robot TWENDY-ONE. In *Proc. of the 2009 IEEE International Conference on Robotics and Automation*, pages 580–586, May 2009

[KD02] Wisama Khalil and Etienne Dombre. *Modeling. Identification & Control of Robots*, Hermes Penton, 2002

[Kel29] Oliver Dimon Kellogg. *Foundations of Potential Theory*. Springer, 1929

[Kha86] Oussama Khatib. Real-Time Obstacle Avoidance for Manipulators and Mobile Robots. *International Journal of Robotics Research*, 5(1):90–98, Spring 1986

[Kha87] Oussama Khatib. A Unified Approach for Motion and Force Control of Robot Manipulators: The Operational Space Formulation. *IEEE Journal of Robotics and Automation*, RA-3(1):43–53, February 1987

[Kha95] Oussama Khatib. Inertial Properties in Robotic Manipulation: An Object-Level Framework. *International Journal of Robotics Research*, 14(1):19–36, February 1995

[Kim13] Melanie Kimmel. Safety and Constraints in Cooperative Human-Robot-Interaction. Master's thesis, Technische Universität München, 2013

[KK88] Laura Kelmar and Pradeep Khosla. Automatic Generation of Kinematics for a Reconfigurable Modular Manipulator System. In *Proc. of the 1988 IEEE International Conference on Robotics and Automation*, pages 663–668, April 1988

[KKK95] Miroslav Krstić, Ioannis Kanellakopoulos, and Petar Kokotović. *Nonlinear and Adaptive Control Design*. Wiley, 1995

[KKM+11] Kenji Kaneko, Fumio Kanehiro, Mitsuharu Morisawa, Kazuhiko Akachi, Go Miyamori, Atsushi Hayashi, and Noriyuki Kanehira. Humanoid Robot HRP-4 - Humanoid Robotics Platform with Lightweight and Slim Body. In *Proc. of the 2011 IEEE/RSJ International Conference on Intelligent Robots and Systems*, pages 4400–4407, September 2011

[KLP13] Leslie Pack Kaelbling and Tomás Lozano-Pérez. Integrated task and motion planning in belief space. *International Journal of Robotics Research*, 32(9-10):1194–1227, August/September 2013

[KLW11] Oussama Kanoun, Florent Lamiraux, and Pierre-Brice Wieber. Kinematic Control of Redundant Manipulators: Generalizing the Task-Priority Framework to Inequality Task. *IEEE Transactions on Robotics* 27(4):785–792, August 2011

[KNK+02] James Kuffner, Koichi Nishiwaki, Satoshi Kagami, Yasuo Kuniyoshi, Masayuki Inaba, and Hirochika Inoue. Self-Collision Detection and Prevention for Humanoid Robots. In *Proc. of the 2002 IEEE International Conference on Robotics and Automation*, pages 2265–2270, May 2002

[KSP08] Oussama Khatib, Luis Sentis, and Jae-Heung Park. A Unified Framework for Whole-Body Humanoid Robot Control with Multiple Constraints and Contacts. In *European Robotics Symposium 2008*, pages 303–312, March 2008

[KSPW04] Oussama Khatib, Luis Sentis, Jaeheung Park, and James Warren. Whole-Body Dynamic Behavior and Control of Human-like Robots. *International Journal of Humanoid Robotics*, 1(1):29–43, March 2004

[LBU09] Sebastian Lohmeier, Thomas Buschmann, and Heinz Ulbrich. Humanoid Robot LOLA. In *Proc. of the 2009 IEEE International Conference on Robotics and Automation*, pages 775–780, May 2009

[LCCF11] Chi-Pang Lam, Chen-Tun Chou, Kuo-Hung Chiang, and Li-Chen Fu. Human-Centered Robot Navigation - Towards a Harmoniously Human-Robot Coexisting Environment. *IEEE Transactions on Robotics*, 27(1):99–112, February 2011

[LDBAS16] Daniel Leidner, Alexander Dietrich, Michael Beetz, and Alin Albu-Schäffer. Knowledge-enabled parameterization of whole-body control strategies for compliant service robots. *Autonomous Robots*, 40(3):519–536, March 2016

[LDS+14] Daniel Leidner, Alexander Dietrich, Florian Schmidt, Christoph Borst, and Alin Albu-Schäffer. Object-Centered Hybrid Reasoning for Whole-Body Mobile Manipulation. In *Proc. of the 2014 IEEE International Conference on Robotics and Automation*, pages 1828–1835, June 2014

[LGDAS14] Dominic Lakatos, Gianluca Garofalo, Alexander Dietrich, and Alin Albu-Schäffer. Jumping Control for Compliantly Actuated Multilegged Robots. In *Proc. of the 2014 IEEE International Conference on Robotics and Automation*, pages 4562–4568, June 2014

[LGP+13] Dominic Lakatos, Martin Görner, Florian Petit, Alexander Dietrich, and Alin Albu-Schäffer. A Modally Adaptive Control for Multi-Contact Cyclic Motions in Compliantly Actuated Robotic Systems: Experiment and Simulation. In *Proc. of the 2013 IEEE/RSJ International Conference on Intelligent Robots and Systems*, pages 5388–5395, November 2013

[Lie77] Alain Liegeois. Automatic Supervisory Control of the Configuration and Behavior of Multibody Mechanisms. *IEEE Transactions on Systems, Man, and Cybernetics*, SMC-7(12):868–871, December 1977

[LMP11] Jaemin Lee, Nicolas Mansard, and Jaeheung Park. Intermediate Desired Value Approach for Continuous Transition among Multiple Tasks of Robots. In *Proc. of the 2011 IEEE International Conference on Robotics and Automation*, pages 1276–1282, May 2011

[LNL+06] Michel Lauria, Isabelle Nadeau, Pierre Lepage, Yan Morin, Patrick Giguère, Fréderic Gagnon, Dominic Létourneau, and François Michaud. Design and Control of a Four Steered Wheeled Mobile Robot. In *Proc. of the IEEE 32nd Annual Conference on Industrial Electronics*, pages 4020–4025, Nov. 2006

[LQXX09] Tin Lun Lam, Huihuan Qian, Yangsheng Xu, and Guoqing Xu. Omni-directional Steer-by-Wire Interface for Four Wheel Independent Steering Vehicle. In *Proc. of the 2009 IEEE International Conference on Robotics and Automation*, pages 1383–1388, May 2009

[LVYK13] Sébastien Lengagne, Joris Vaillant, Eiichi Yoshida, and Abderrahmane Kheddar. Generation of whole-body optimal dynamic multi-contact motions. *International Journal of Robotics Research*, 32(9–10):1104–1119, August/September 2013

[Man12] Nicolas Mansard. A Dedicated Solver for Fast Operational-Space Inverse Dynamics. In *Proc. of the 2012 IEEE International Conference on Robotics and Automation*, pages 4943–4949, May 2012

[MB11] Lorenz Mösenlechner and Michael Beetz. Parameterizing Actions to have the Appropriate Effects. In *Proc. of the 2011 IEEE/RSJ International Conference on Intelligent Robots and Systems*, pages 4141–4147, September 2011

[MCR96] Éric Marchand, François Chaumette, and Alessandro Rizzo. Using the task function approach to avoid robot joint limits and kinematic singularities in visual servoing. In *Proc. of the 1996 IEEE/RSJ International Conference on Intelligent Robots and Systems*, pages 1083–1090, November 1996

[MGG+13] Federico L. Moro, Michael Gienger, Ambarish Goswami, Nikos G. Tsagarakis, and Darwin G. Caldwell. An Attractor-based Whole-Body Motion Control (WBMC) System for Humanoid Robots. In *Proc. of the 13th IEEE-RAS International Conference on Humanoid Robots*, pages 42–49, October 2013

[MK89] Anthony A. Maciejewski and Charles A. Klein. The Singular Value Decomposition: Computation and Application to Robotics. *International Journal of Robotics Research*, 8(6):63–79, December 1989

[MK08] Nicolas Mansard and Oussama Khatib. Continuous Control Law from Unilateral Constraints: Application to Reactive Obstacle Avoidance in Operational Space. In *Proc. of the 2008 IEEE International Conference on Robotics and Automation*, pages 3359–3364, May 2008

[MKK09] Nicolas Mansard, Oussama Khatib, and Abderrahmane Kheddar. A Unified Approach to Integrate Unilateral Constraints in the Stack of Tasks. *IEEE Transactions on Robotics*, 25(3):670–685, June 2009

[MLS94] Richard M. Murray, Zexiang Li, and S. Shankar Sastry. *A Mathematical Introduction to Robotic Manipulation*. CRC Press, 1994

[MRC09] Nicolas Mansard, Oussama Khatib, and Abderrahmane Kheddar. Continuity of Varying-Feature-Set Control Laws. *IEEE Transactions on Automatic Control*, 54(11):2493–2505, November 2009

[NCM+08] Jun Nakanishi, Rick Cory, Michael Mistry, Jan Peters, and Stefan Schaal. Operational Space Control: A Theoretical and Empirical Comparison. *International Journal of Robotics Research*, 27(6):737–757, June 2008

[NH86] Yoshihiko Nakamura and Hideo Hanafusa. Inverse Kinematic Solutions With Singularity Robustness for Robot Manipulator Control. *Journal of Dynamic Systems, Measurement, and Control*, 108(3):163–171, September 1986

[NHY87] Yoshihiko Nakamura, Hideo Hanafusa, and Tsuneo Yoshikawa. Task-Priority Based Redundancy Control of Robot Manipulators. *International Journal of Robotics Research* 6(2):3–15, June 1987

[NKS+10] Ken'ichiro Nagasaka, Yasunori Kawanami, Satoru Shimizu, Takashi Kito, Toshimitsu Tsuboi, Atsushi Miyamoto, Tetsuharu Fukushima, and Hideki Shimomura. Whole-body Cooperative Force Control for a Two-Armed and Two-Wheeled Mobile Robot Using Generalized Inverse Dynamics and Idealized Joint Units. In *Proc. of the 2010 IEEE International Conference on Robotics and Automation*, pages 3377–3383, May 2010

[NSV99] Ciro Natale, Bruno Siciliano, and Luigi Villani. Spatial Impedance Control of Redundant Manipulators. In *Proc. of the 1999 IEEE International Conference on Robotics and Automation*, pages 1788–1793, May 1999

[OASK+04] Christian Ott, Alin Albu-Schäffer, Andreas Kugi, Stefano Stramigioli, and Gerd Hirzinger. A Passivity Based Cartesian Impedance Controller for Flexible Joint Robots - Part I: Torque Feedback and Gravity Compensation. In *Proc. of the 2004 IEEE International Conference on Robotics and Automation*, pages 2659–2665, April 2004

[OASKH08] Christian Ott, Alin Albu-Schäffer, Andreas Kugi, and Gerd Hirzinger. On the Passivity-Based Impedance Control of Flexible Joint Robots. *IEEE Transactions on Robotics*, 24(2):416–429, April 2008

[OCY98] Yonghwan Oh, Wankyun Chung, and Youngil Youm. Extended Impedance Control of Redundant Manipulators Based on Weighted Decomposition of Joint Space. *Journal of Robotic Systems*, 15(5): 231–258, May 1998

[ODAS15] Christian Ott, Alexander Dietrich, and Alin Albu-Schäffer. Prioritized Multi-Task Compliance Control of Redundant Manipulators. *Automatica*, 53:416–423, March 2015

[ODR14] Christian Ott, Alexander Dietrich, and Maximo A. Roa. Torque-based multi-task and balancing control for humanoid robots. In *Proc. of the 11th International Conference on Ubiquituous Robots and Ambient Intelligence*, pages 143–144, November 2014

[OEF+06] Christian Ott, Oliver Eiberger, Werner Friedl, Berthold Bäuml, Ulrich Hillenbrand, Christoph Borst, Alin Albu-Schäffer, Bernhard Brunner, Heiko Hirschmüller, Simon Kielhöfer, Rainer Konietschke, Michael Suppa, Thomas Wimböck, Franziska Zacharias, and Gerd Hirzinger. A Humanoid Two-Arm System for Dexterous Manipulation. In *Proc. of the 6th IEEE/RAS International Conference on Humanoid Robots*, pages 276–283, December 2006

[OHL13] Christian Ott, Bernd Henze, and Dongheui Lee. Kinesthetic teaching of humanoid motion based on whole-body compliance control with interaction-aware balancing. In *Proc. of the 2013 IEEE/RSJ International Conference on Intelligent Robots and Systems*, pages 4615–4621, November 2013

[OKN08] Christian Ott, Andreas Kugi, and Yoshihiko Nakamura. Resolving the Problem of Non-integrability of Nullspace Velocities for Compliance Control of Redundant Manipulators by using Semi-definite Lyapunov functions. In *Proc. of the 2008 IEEE International Conference on Robotics and Automation*, pages 1999–2004, May 2008

[Ott08] Christian Ott. *Cartesian Impedance Control of Redundant and Flexible-Joint Robots. Springer Tracts in Advanced Robotics*, vol. 49. Springer Publishing Company, Berlin Heidelberg, 2008

[PA93] Zhi-Xin Peng and Norihiko Adachi. Compliant Motion Control of Kinematically Redundant Manipulators. *IEEE Transactions on Robotics and Automation* 9(6):831–837, December 1993

[Par99] Jonghoon Park. *Analysis and Control of Kinematically Redundant Manipulators: An Approach Based on Kinematically Decoupled Joint Space Decomposition*. PhD thesis, Pohang University of Science and Technology, 1999

[Pau83] Richard P. Paul. *Robot Manipulators: Mathematics, Programming, and Control.* The MIT Press, 1983

[PAW10] Robert Platt, Muhammad Abdallah, and Charles Wampler. Multi-Priority Cartesian Impedance Control. In *Proc. of Robotics, Science and Systems*, June 2010

[PAW11] Robert Platt, Muhammad Abdallah, and Charles Wampler. Multiple-priority impedance control. In *Proc. of the 2011 IEEE International Conference on Robotics and Automation*, pages 6033–6038, May 2011

[PCY99] Jonghoon Park, Wankyun Chung, and Youngil Youm. On Dynamical Decoupling of Kinematically Redundant Manipulators. In *Proc. of the 1999 IEEE/RSJ International Conference on Intelligent Robots and Systems*, pages 1495–1500, October 1999

[PDAS15] Florian Petit, Alexander Dietrich, Alin Albu-Schäffer. Generalizing Torque Control Concepts: Using Well-Established Torque Control Methods on Variable Stiffness Robots. *IEEE Robotics & Automation Magazine*, 22(4):37–51, December 2015

[PMU+08] Jan Peters, Michael Mistry, Firdaus Udwadia, Jun Nakanishi, and Stefan Schaal. A unifying framework for robot control with redundant DOFs. *Autonomous Robots*, 24(1): 1–12, 2008

[PSK11] Roland Philippsen, Luis Sentis, and Oussama Khatib. An Open Source Extensible Software Package to Create Whole-Body Compliant Skills in Personal Mobile Manipulators. In *Proc. of the 2011 IEEE/RSJ International Conference on Intelligent Robots and Systems*, pages 1036–1041, September 2011

[Rau13] Peter Rausch. Design of a Cartesian Impedance Controller for a Humanoid Robot with Significant Passive Compliance. Master's thesis, Technische Universität München, 2013

[RDC14] Jens Reinecke, Alexander Dietrich, and Maxime Chalon. Experimental Comparison of Slip Detection Strategies by Tactile Sensing with the BioTac on the DLR Hand Arm System. In *Proc. of the 2014 IEEE International Conference on Robotics and Automation*, pages 2742–2748, June 2014

[SEM+08] Olivier Stasse, Adrien Escande, Nicolas Mansard, Sylvain Miossec, Paul Evrard, and Abderrahmane Kheddar. Real-Time (Self)-Collision Avoidance Task on a HRP-2 Humanoid Robot. In *Proc. of the 2008 IEEE International Conference on Robotics and Automation*, pages 3200–3205, May 2008

[SGJG07] Hisashi Sugiura, Michael Gienger, Herbert Janssen, and Christian Goerick. Real-Time Collision Avoidance with Whole Body Motion Control for Humanoid Robots. In *Proc. of the 2007 IEEE/RSJ International Conference on Intelligent Robots and Systems*, pages 2053–2058, October 2007

[SGJG10] Hisashi Sugiura, Michael Gienger, Herbert Janssen, and Christian Goerick. Reactive Self Collision Avoidance with Dynamic Task Prioritization for Humanoid Robots. *International Journal of Humanoid Robotics*, 7(1):31–54, 2010

[SJK97] R. Sepulchre, M. Jankovic, and P. Kokotovic. *Constructive Nonlinear Control*. Springer, 1997

[SK05] Luis Sentis and Oussama Khatib. Synthesis of Whole-Body Behaviors through Hierarchical Control of Behavioral Primitives. *International Journal of Humanoid Robotics*, 2(4):505–518, January 2005

[SK08] Bruno Siciliano and Oussama Khatib. *Springer Handbook of Robotics*. Springer, 2008

[SKVS12] Hamid Sadeghian, Mehdi Keshmiri, Luigi Villani, and Bruno Siciliano. Priority Oriented Adaptive Control of Kinematically Redundant Manipulators. In *Proc. of the 2012 IEEE International Conference on Robotics and Automation*, pages 293–298, May 2012

[SOG10] Mike Stilman, Jon Olson, and William Gloss. Golem Krang: Dynamically Stable Humanoid Robot for Mobile Manipulation. In *Proc. of the 2010 IEEE International Conference on Robotics and Automation*, pages 3304–3309, May 2010

[SPK10] Luis Sentis, Jaeheung Park, and Oussama Khatib. Compliant Control of Multicontact and Center-of-Mass Behavior in Humanoid Robots. *IEEE Transactions on Robotics* 26(3):483–501, June 2010

[SPP13] Luis Sentis, Josh Petersen, and Roland Philippsen. Implementation and stability analysis of prioritized whole-body compliant controllers on a wheeled humanoid robot in uneven terrains. *Autonomous Robots*, 35(4):301–319, 2013

[SRK+13] Layale Saab, Oscar E. Ramos, François Keith, Nicolas Mansard, Philippe Souères, and Jean-Ives Fourquet. Dynamic Whole-Body Motion Generation Under Rigid Contacts and Other Unilateral Constraints. *IEEE Transactions on Robotics*, 29(2):346–362, April 2013

[SS91] Bruno Siciliano and Jean-Jacques Slotine. A General Framework for Managing Multiple Tasks in Highly Redundant Robotic Systems. In *Proc. of the 5th International Conference on Advanced Robotics*, pages 1211–1216, June 1991

[SVKS13] Hamid Sadeghian, Luigi Villani, Mehdi Keshmiri, and Bruno Siciliano. Dynamic multi-priority control in redundant robotic systems. *Robotica*, 31(7):1155–1167, October 2013

[SWA+02] Yoshiaki Sakagami, Ryujin Watanabe, Chiaki Aoyama, Shinichi Matsunaga, Nobuo Higaki, and Kikuo Fujimura. The intelligent ASIMO: System overview and integration. In *Proc. of the 2002 IEEE/RSJ International Conference on Intelligent Robots and Systems*, pages 2478–2483, October 2002

[TBF11] Holger Täubig, Berthold Bäuml, and Udo Frese. Real-time Swept Volume and Distance Computation for Self Collision Detection. In *Proc. of the 2011 IEEE/RSJ International Conference on Intelligent Robots and Systems*, pages 1585–1592, September 2011

[TDNM96] Benoit Thuilot, Brigitte D'Andréa-Novel, and Alain Micaelli. Modeling and Feedback Control of Mobile Robots Equipped with Several Steering Wheels. *IEEE Transactions on Robotics and Automation* 12(3):375–390, June 1996

[UASvdS04] Holger Urbanek, Alin Albu-Schäffer, and Patrick van der Smagt. Learning from Demonstration: Repetitive Movements for Autonomous Service Robotics. In *Proc. of the 2004 IEEE/RSJ International Conference on Intelligent Robots and Systems*, pages 3495–3500, September 2004

[vdS00] Arjan van der Schaft. *L_2-Gain and Passivity Techniques in Nonlinear Control*. Springer Publishing Company, Berlin Heidelberg, 2nd edition, 2000

[WA99] Masayoshi Wada and Haruhiko H. Asada. Design and Control of a Variable Footprint Mechanism for Holonomic Omnidirectional Vehicles and its Application to Wheelchairs. *IEEE Transactions on Robotics and Automation*, 15(6):978–989, December 1999

[Wam86] Charles W. Wampler. Manipulator Inverse Kinematic Solutions Based on Vector Formulations and Damped Least-Squares Methods. *IEEE Transactions on Systems, Man, and Cybernetics*, 16(1):93–101, January 1986

[Wim12] Thomas Wimböck. *Controllers for Compliant Two-Handed Dexterous Manipulation*. PhD thesis, Vienna University of Technology, December 2012

[WMR10] Jason Wolfe, Bhaskara Marthi, and Stuart Russell. Combined Task and Motion Planning for Mobile Manipulation. In *Proc. of the 20th International Conference on Automated Planning and Scheduling*, pages 254–257, May 2010

[WO12] Thomas Wimböck and Christian Ott. Dual-Arm Manipulation. In *Towards Service Robots for Everyday Environments*. volume 76 of *Springer Tracts in Advanced Robotics*, pages 353–366. Springer, Berlin Heidelberg, 2012

[WOH07] Thomas Wimböck, Christian Ott, and Gerd Hirzinger. Impedance Behaviors for Two-handed Manipulation: Design and Experiments. In *Proc. of the 2007 IEEE International Conference on Robotics and Automation*, pages 4182–4189, April 2007

[WSA11] Kyle N. Winfree, Paul Stegall, and Sunil K. Agrawal. Design of a Minimally Constraining, Passively Supported Gait Training Exoskeleton: ALEX II. In *Proc. of the 2011 IEEE International Conference on Rehabilitation Robotics*, pages 1–6, June 2011

[YK09] Taizo Yoshikawa and Oussama Khatib. Compliant Humanoid Robot Control by the Torque Transformers. In *Proc. of the 2009 IEEE/RSJ International Conference on Intelligent Robots and Systems*, pages 3011–3018, October 2009

[Yos90] Tsuneo Yoshikawa. *Foundations of Robotics: Analysis and Control*. The MIT Press, Cambridge, 1990

[ZDWS04] Erkan Zergeroglu, Darren D. Dawson, Ian W. Walker, and Pradeep Setlur. Nonlinear Tracking Control of Kinematically Redundant Robot Manipulators. *IEEE/ASME Transactions on Mechatronics*, 9(1):129–132, March 2004

Printed in the United States
By Bookmasters